THE DESIGN AND DESCRIPTION OF COMPUTER ARCHITECTURES

THE DESIGN AND DESCRIPTION OF COMPUTER ARCHITECTURES

SUBRATA DASGUPTA

University of Southwestern Louisiana
Lafayette, Louisiana

A Wiley-Interscience Publication
JOHN WILEY & SONS

New York　　　Chichester　　　Brisbane　　　Toronto　　　Singapore

UNIX is a trademark of Bell Laboratories.

ADA is a registered trademark of the U.S. Government ADA Joint Program Office.

Copyright © 1984 by John Wiley & Sons, Inc.

All rights reserved. Published simultaneously in Canada.

Reproduction or translation of any part of this work beyond that permitted by Section 107 or 108 of the 1976 United States Copyright Act without the permission of the copyright owner is unlawful. Requests for permission or further information should be addressed to the Permissions of Department, John Wiley & Sons, Inc.

Library of Congress Cataloging in Publication Data:

Dasgupta, Subrata.
   The design and description of computer architectures.

   "A Wiley-Interscience publication."
   Bibliography: p.
   Includes index.
   1. Computer architecture. 2. System design.
I. Title.

QA76.9.A73D36   1984   621.3819'58   83-21826
ISBN 0-471-89616-0

Printed in the United States of America

10 9 8 7 6 5 4 3 2 1

TO
MY PARENTS
AND
MY PARENTS-IN-LAW

PREFACE

The design of computer architectures, an aspect of the broader discipline of computer design, has traditionally suffered from two major drawbacks. First, the idea of storing the architectural design in some formal representation has remained largely outside the mainstream of architectural thought and practice; second, in the absence of a theoretical or formal framework, architectural designs have conventionally been evaluated with respect to both logical correctness and performance only after they have been implemented as physical systems.

These are major shortcomings for any field that lays some claim to being called a discipline. Their seriousness will be evident at one time or another to all those involved with computer architecture, whether as a designer, teacher, or theorist. For their consequences are a compendium of undesirable characteristics that are commonly encountered by all sectors of the architectural community:

1. An inability to communicate the design in a precise, unambiguous way.
2. The invariable presence in the design of features that are undefined and hence subject to erroneous interpretation.
3. An inability to understand the architecture, both from a systems point of view and at the level of minute detail.
4. A low level of confidence in the logical correctness and reliability of the design.
5. A basic inability to manipulate the design either for the purpose of exploring alternative choices for some part of the architecture or with the intent of extending or modifying an existing design.

Indeed, some or all items from this list will appear familiar to all those who are involved in the design of complex systems, such as the software designer, urban planner, or (civil/structural) architect. In the area of software design, many of these characteristics were symptomatic of a fundamental weakness in our understanding of the nature and complexity of software in the late 1960s. This weakness prompted the development of a series of techniques, concepts, and ideas, largely pioneered by E. W. Dijkstra, C. A. R. Hoare, N. Wirth, and others, that in recent times have nucleated into a discipline dubbed "software engineering." In urban and environmental planning and architecture, the response to this problem of unmastered complexity is seen in the form of a plea for rational design, even a mathematization of design, notably exemplified by Christopher Alexander's influential monograph *Notes on the Synthesis of Form* (1964). More generally and in some ways more interestingly, with the recurring observation of the same problem of design in diverse fields, there has emerged from the 1960s onward a vigorous group of design theorists whose focus is the design process itself. For computer scientists, the pioneering and still most illuminating work in this respect must remain Herbert Simon's beautiful little book *The Sciences of the Artificial* (1981).

Computer architecture does not yet, apparently, face the kind of crisis that confronts software design or urban planning. Thus, the attempt to develop a theory or methodology of computer architecture design is less motivated by any perceived crises in this field than by a deep sense of dissatisfaction (on the part of this writer, at any rate) with the state of a design discipline that is ridden with flaws.

There are, however, indications that because of startling advances in technology, computer architects may well in the very near future face the problem of unmastered complexity so familiar to software designers. For instance, the development of large-scale integration (LSI) and (now) very large-scale integration (VLSI) semiconductor technology has resulted in a potential for processor chips of unprecedented gate complexity. Thus computer architects are already grappling with design problems that are quite different from those of a decade ago, simply because the technological constraints have changed so rapidly (Treleaven [1980]).

Furthermore, as an outcome of the large-scale availability of cost-effective microprocessors, the possibilities of distributed systems, networks, and multiprocessors are now widely recognized. Thus computer structures of unprecedented organizational complexity are being studied, designed, and experimented with.

Finally, these same technological changes have made the vertical migration of operating system functions and language primitives down to the microcode level a viable system implementation option. Thus microprograms are likely to become progressively larger and more complex, with an attendant increase in functionality and complexity of the overall architecture.

We are thus well advised to anticipate the emergence of far greater complexity in the architectural domain than might seem to exist at present.

PREFACE

This monograph is about the methodology of design of computer architecture. The term "methodology" is used here in its narrowest sense of "the study or description of the methods or procedures used in some activity" (Bullock and Stallybrass [1977], pp. 387–388)—in this case, computer architecture. In particular, this book is concerned with a particular approach to the design of architectures, based on the use of symbolic descriptions.

Style, according to H. A. Simon, is one way of doing things chosen from a number of alternative ways. It is normally identified by characteristics of the objects designed—as, for example, when we talk about a given computer processing a pipeline style of architecture on the basis of certain characteristics of information flow through the hardware during the course of a computation. As Simon has pointed out, it may also arise from differences in the design process. A good example from the software domain is provided by D. L. Parnas (1972) who, in a famous paper, showed that depending on the criteria used for decomposing systems into modules, one may obtain a software product organized in two quite different ways. The design process leaves its imprint on the final product; the alternative modularization techniques represent distinct design styles.

From this viewpoint it may be fair to say that the focus here is also on a particular style of architectural design, a style characterized mainly by the use of symbolic description languages for the specification of architecture at different, significant levels of abstraction.

Needless to say, it is not my place to advocate this approach to thinking about and designing computer architectures as *the* style to adopt. It is my objective, however, to promulgate the view that a sine qua non for the emergence of computer architecture as a design discipline is the elimination of the flaws described at the beginning of this preface. With a symbolic description language, the activities of designing, evaluating, modifying, and understanding an architecture can be conducted quite apart from, largely independent of, and preferably prior to the task of implementing the architecture in physical hardware. The proposed design style, as well as several other issues of architectural design discussed in this monograph, is intended as a contribution to such a discipline of computer architecture.

A few words must be said about the languages defined and used in this book. These languages (which, for reasons that will be evident later, form part of what I call the S* family) represent the outcome of work carried out over the past few years by this author and his students, on the development of a *family* of languages for the design, description, and verification of computer architectures and microprograms. Indeed, the present monograph is largely the culmination of this earlier work. Thus, a detailed discussion of the S* family as part of the present subject matter seemed both natural and necessary.

At the same time I must add a disclaimer. This monograph must not be interpreted as advocating these particular languages in preference to others that have been proposed in the literature. The rationale underlying the design of the S* family has been stated in earlier publications and is repeated here for the sake of completeness. It must remain for the reader to assess the family and decide on its usefulness relative to alternative proposals. Needless to say, as the history of programming languages amply illustrates, much experience must be gained with the application of design languages before any proper assessment can be achieved. At this time of writing, probably the only architecture description (or microprogramming) language that has been adopted to any degree by users outside the immediate locus of its invention is ISPS (Barbacci et al. [1978]). One of the reasons for devoting considerable space in this book to a detailed discussion of the S* family is the hope that others will seek to apply it, criticize it, and provide the necessary insight for the derivation of better and more elegant languages in the future.

SUBRATA DASGUPTA

Lafayette, Louisiana
November 11, 1983

ACKNOWLEDGMENTS

I am grateful for having had the opportunity of presenting some of the ideas discussed in this book in seminars at various universities and conferences over the last two years.

Parts of the manuscript were read by Werner Damm, Alan Wagner, Joseph Linn, Prasenjit Biswas, Harold Lorin, and Marius Olafsson. I thank them for their comments. I am also indebted to Marius Olafsson for allowing me to quote excerpts from his thesis.

At one time or another, I have had discussions with many colleagues and students on the topics of computer architecture, microprogramming, verification, and design methodology, many of which have influenced the shape of this book. In particular, I remember with pleasure conversations with Maurice Wilkes, Werner Damm, Michael Flynn, Alan Wagner, Marius Olafsson, Bruce Shriver, and Joseph Linn. Needless to say, none of the above are responsible for the opinions and ideas expressed in this book.

I am extremely grateful to Judith Abbott, who typed the major part of the book and to Cathy Pomier and Phil Wilsey for their assistance in completing the preparation of the manuscript. I must also thank James Gaughan and the staff of Wiley-Interscience for their invaluable editorial help.

My thanks to the Institute of Electrical and Electronics Engineers, the Association for Computing Machinery, AFIPS Press, John Wiley and Sons, Academic Press, and Springer Verlag for granting me permission to reproduce diagrams and excerpts from their publications.

Finally, a very personal note of gratitude to my parents, my parents-in-law, and to my wife for their love, support, and understanding over the years.

S.D.

CONTENTS

CHAPTER 1 WHAT IS COMPUTER ARCHITECTURE?   1

   1.1 Introduction 1
   1.2 Levels of Architecture 4
   1.3 Conceptual Integrity of Architectures 8
   1.4 Bibliographic Remarks 10

CHAPTER 2 THE INFORMAL DESIGN PROCESS   11

   2.1 Architecture as Craft 11
   2.2 Bibliographic Remarks 16

CHAPTER 3 THE FORMAL DESIGN PROCESS   18

   3.1 Introduction 18
   3.2 What Is Methodology? 19
   3.3 The Limits of Formal Design 22
   3.4 Bibliographic Remarks 23

CHAPTER 4 ISSUES IN LANGUAGE DESIGN   24

   4.1 Introduction 24
   4.2 Levels of Abstraction 25
   4.3 The Operational-Functional Dichotomy 28
   4.4 Procedural and Nonprocedural Descriptions 30
   4.5 Structure and Behavior 32
   4.6 The Influence of Programming Languages 37
   4.7 Bibliographic Notes 38

CHAPTER 5 A LANGUAGE FOR DESCRIBING COMPUTER ARCHITECTURES — 40

5.1 Toward a Family of Design Languages 40
5.2 An Overview of the S* Family 41
5.3 Data Types and Data Objects in S*A 43
5.4 The Specification of Action in Architectural Descriptions 46
5.5 Modular Descriptions 48
5.6 Mechanisms and Systems 50
5.7 Asynchronous Concurrent Systems 55
5.8 Synonyms 57
5.9 Mechanism Types 58
5.10 Bibliographic Notes 61

CHAPTER 6 FORMAL DESIGN OF A MICROCODE LOADER — 62

6.1 The Problem 62
6.2 An Informal Design 64
6.3 A Formal Description 66
6.4 A Diversion: Floyd-Hoare Correctness Proofs 70
6.5 Derivation of a Correct Mechanism 73
6.6 Bibliographic Notes 84

CHAPTER 7 AN ASYNCHRONOUS ARCHITECTURAL SYSTEM — 85

7.1 Overview of the System 85
7.2 Formal Description 86
7.3 Bibliographic Notes 91

CHAPTER 8 ON THE CORRECTNESS OF ASYNCHRONOUS ARCHITECTURAL SYSTEMS — 92

8.1 Introduction 92
8.2 Principles of the Owicki-Gries Technique 93
8.3 Application to Architectural Verification 95
8.4 Verification of DATAFLOW 98
8.5 Bibliographic Remarks 109

CHAPTER 9 TOWARD HIGH-LEVEL MICROPROGRAMMING — 110

9.1 Problems of High-Level Microprogramming 111
9.2 Some Approaches to High-Level Microprogramming 119
9.3 Bibliographic Remarks 125

CONTENTS

CHAPTER 10 A MICROPROGRAMMING LANGUAGE SCHEMA   127

- 10.1 Data Types and Data Objects 127
- 10.2 Synonyms 131
- 10.3 Executional Statements 132
- 10.4 Bibliographic Remarks 139

CHAPTER 11 A PRACTICAL MICROPROGRAMMING LANGUAGE   140

- 11.1 Introduction 140
- 11.2 Architecture of the QM-1 142
- 11.3 Description of S*(QM-1) 158
- 11.4 A Programming Example in S*(QM-1) 170
- 11.5 The Significance of S*(QM-1) 171
- 11.6 Bibliographic and Other Remarks 172

CHAPTER 12 ON STYLE IN COMPUTER ARCHITECTURE   174

- 12.1 Introduction 174
- 12.2 Style Induced by Architectural Features 176
- 12.3 The Influence of Architectural Style on the Design Process 180
- 12.4 The Influence of Implementation Style 184
- 12.5 Architectural Design Style 185
- 12.6 Bibliographic Remarks 190

CHAPTER 13 THE OUTER ENVIRONMENT   192

- 13.1 Introduction 192
- 13.2 Ordering Decisions in Exoarchitectural Design 193
- 13.3 Characterizing the Outer Environment 197
- 13.4 Other Aspects of the Outer Environment 210

CHAPTER 14 DESIGN OF AN ARCHITECTURE: A CASE STUDY   212

- 14.1 Toward a Discipline of Computer Architecture 212
- 14.2 Two Points to Ponder 214
- 14.3 Design of the QM-C Architecture 216

EPILOGUE   246

APPENDIX A A DEFINITION OF THE ARCHITECTURAL DESCRIPTION LANGUAGE S*A   249

APPENDIX B	SYNTAX AND SEMANTICS OF THE MICROPROGRAMMING LANGUAGE SCHEMA S*	275
REFERENCES		281
INDEX		293

THE DESIGN AND DESCRIPTION OF COMPUTER ARCHITECTURES

ONE

WHAT IS COMPUTER ARCHITECTURE?

1.1 INTRODUCTION

Precisely what constitutes *computer architecture* has been a point of some debate in computer science. The crux of the problem appears to be the complex, hierarchic nature of computers. They are, first of all, complex in Simon's (1981) sense—they are composed of a *large number of parts* that interact in a nontrivial way; and hierarchy is a necessary way of organizing such a complex system. This is illustrated schematically by Fig. 1.1, where the individual system parts a through g are conceptually, logically, or physically (depending on the actual system) organized into two higher hierarchic levels. The relationship between adjacent levels is one of "consists of." Thus, α consists of A, B, and C; A consists of a, b, c, and so on.

In the case of computer systems, this kind of hierarchy is seen in the way we divide information into chunks when we design, describe, or understand such systems; thus, we talk of a system being composed of a processor, a memory, and a control unit. The processor in turn may consist of a local store, an arithmetic logic unit (ALU), multiplexers, and buses, while the control unit may be composed of a control memory, a set of specialized registers and a sequencer, and so on (Fig. 1.2).

A second kind of complexity arises from the many different *levels of description* that exist for computers. As pointed out by Bell and Newell (1971), these levels are not equivalent in the sense that anything said one way can be said another. On the contrary, each description level performs a

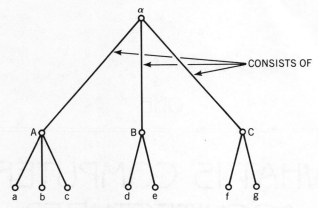

Figure 1.1 Hierarchy of several interacting components.

function that cannot be performed adequately by another. This is illustrated schematically by Fig. 1.3. The essential relationship between levels is one of abstraction or, conversely, refinement. Thus, level 1 is an abstraction of level 2; conversely, level 2 is a refinement of level 1.

A third kind of complexity arises when a system is *constructed* level by level. System complexity in this context appears—at least in some cases—to be the effect rather than the cause of multiple construction levels, since a system may have been designed or constructed in this form for reasons having nothing to do intrinsically with the management of complexity. The resulting system appears complex as a consequence of its many levels. Note that the relationship between levels here is one of "is implemented on" or "is constructed on" (Fig. 1.4).

A computer system, then, is complex and hierarchic in a number of different ways: it is composed of a large number of parts that interact in a nontrivial way; it is rich in the variety of levels at which it may be meaningfully described; and it may possibly be constructed as a many-layered system.

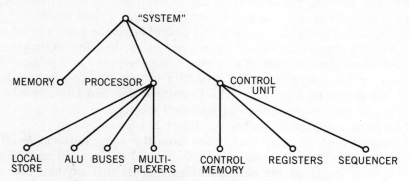

Figure 1.2 Hierarchy in a computer hardware system.

INTRODUCTION

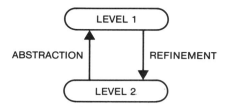

Figure 1.3 Hierarchy due to several levels of description.

Example 1.1

Figure 1.5 shows schematically a hierarchic, multilevel view of the CAP system, a machine that offers its users a protected memory system based on the *capability principle* (Wilkes and Needham [1980]). The user's view of the system is represented in this diagram by the level 3 CAP system, which is constructed upon the next lower level by means of the CAP operating system.

Roughly speaking, the level 2 system is characterized by a set of functional specifications that the systems programmer must know in order to write the operating system—it defines the CAP machine as seen by the operating systems designer. The level itself is an abstraction of a host of other characteristics that need not (indeed, should not) be visible at this level, for example, the precise nature of the instruction fetch/execute cycle, or the mechanism whereby a capability—an entity that specifies the nature of the access rights a given process has to a particular memory segment—is evaluated.

There must, of course, exist some level in the CAP system at which these characteristics are visible; this is designated as level 1 in Fig. 1.5. The distinction between levels 1 and 2 is one of *information hiding* (a concept originally developed by Parnas [1972]), in the sense that many of the design decisions that are visible in the level 1 system are deliberately made opaque at level 2. In fact, a similar distinction exists between levels 2 and 3: at the

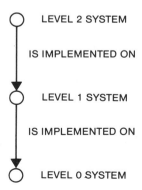

Figure 1.4 Hierarchy due to several levels of construction.

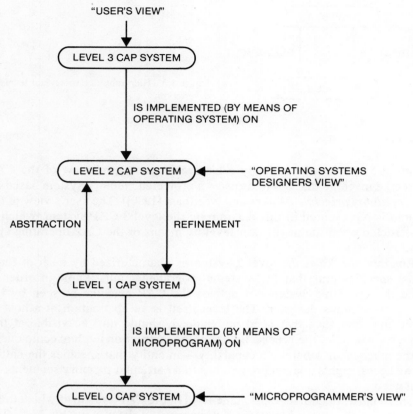

Figure 1.5 Hierarchy in the CAP system.

higher of these levels, the distribution of functions across the software/hardware-firmware interface is not visible, since it may not be clear just which of the machine functions available to the user are implemented by the operating system and which by the basic machine. This information is hidden from the user.

Extending this notion to the lowest pair of levels in the diagram, the distribution of functions across the firmware/hardware interface is hidden in the level 1 system but visible at level 0. The latter characterizes the machine as seen by the microprogrammer.

1.2 LEVELS OF ARCHITECTURE

It must be noted that I am not suggesting that the CAP designers consciously designed their system in the manner just outlined. This system was simply selected as an instance of a common, hierarchic, multilevel structure that is manifested, with minor variations, by all computer systems. In addition it provides clear examples of architectural levels, the multiplicity of which adds to the problem of architectural design.

In this section I shall present a characterization of computer architecture based on the notion of levels of description. While this characterization offers no startling new insight it does attempt, on the one hand, to unify several different definitions of computer architecture proposed in the past, and to resolve, on the other, the contradictions between some of these definitions and the social practice of architectural design, research, and teaching.

One of the most common views of architecture is exemplified by Fuller et al. (1977a), the authors of the reports on the computer family architecture (CFA) project, who used the term to mean *the structure of the computer the programmer needs to know in order to write correct machine language programs*. The authors, following the distinction first made by the IBM System/360 designers, characterize architecture essentially as the interface the hardware presents to the programmer (level 2 in Fig. 1.5).

A similar viewpoint is presented by Myers (1982, p. 6), when he states that:

> computer architecture is the abstraction or definition of a physical system (microcode and hardware) as seen by a machine language programmer or a compiler writer. It is the definition of the conceptual structure and functional behavior of a processor as opposed to such factors as the processor's underlying data flow and controls, logic design and circuit technology.

Myers further specifies that design or decision problems of computer architecture are essentially threefold: the representation of programs, the schemes for data addressing, and the representation of data.

An important (and original) distinction that Myers makes is his recognition of the existence of *several kinds of architecture* within the physical system. What he terms computer architecture (in contrast to other architecture types such as processor or microprogram architecture) is simply one of these kinds.

Since levels of architecture constitute a major theme of this monograph, I shall return to them shortly. But it is instructive to compare the above definitions with some others. Thus, Reddi and Feustal (1976) suggested a framework for describing architectures that includes physical organization, control and flow of information, and the representation, interpretation, and transformation of information. And in a recent review of the subject, Dennis et al. (1979) consider the architecture of a computer system as defining both the hardware/software interface and the means by which this interface is realized by subunits of the computer system.

Clearly, in these last two views, the internal structure of the system appears as an aspect of computer architecture. A similar point is made by Baer (1980), who in defining the scope of the subject as the design of an integrated system that provides a useful tool for the programmer, makes special mention of the structure of the internal components of the system as well as their interconnections, parallel activities, and cooperation.

An examination of the actual topics computer architects concern themselves with, in the spheres of both design and research, indicates that this characterization is far more descriptive of reality than the earlier view. That is, architects are in fact as much concerned with issues related to the internal structure of the computer as they are with its external characteristics. Yet it is clear that the problems of architectural design are not simply those of hardware design. In fact, simply basing our observation on the social practice of architecture, we may identify the following general characteristics:

1. An architecture is an abstraction of the hardware in that it is concerned with the structure and behavior of hardware, represented as an abstract information processing system (rather than as an ensemble of physico-electronic devices).
2. Architectural attributes include both the external (i.e., functional) appearance of the computer as well as its internal form.

Consequently, the computer architect is a designer of information processing systems of a particular kind: those that are directly realized by a combination of hardware and microcode.

Point 2 above implies once more the notion of architectural levels. We may give more complete shape to this notion as follows:

EXOARCHITECTURE. A computer's exoarchitecture is the logical structure and functional capabilities of the hardware system as visible to the machine language programmer or compiler writer.[1]

ENDOARCHITECTURE. A computer's endoarchitecture consists of a specification of the functional capabilities of its physical components, the logical structure of their interconnections, the nature of the information flow between components, and the means whereby this flow of information is controlled.

To understand further the relationship between exoarchitecture and endoarchitecture, it may be noted that a major function of the former is to hide certain kinds of information concerning the computer's design. These include, for example, whether or not the instruction cycle is pipelined, the presence or absence of a cache memory, whether memory interleaving is

---

[1] *Exoarchitecture* then refers to a level of machine abstraction that is also referenced by such terms as *instruction set architecture* or the *instruction set processor* (ISP) level.

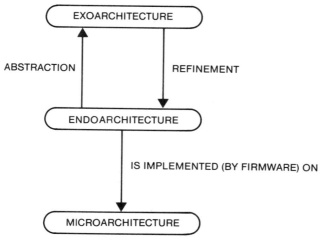

Figure 1.6 Architectural levels.

used, whether instruction interpretation is done in firmware or by hardware, and so on. These are, all, typical endoarchitectural features. Thus, *exoarchitecture is an abstraction of endoarchitecture*. Conversely, the latter may be viewed as what is revealed of the machine's internal logical structure and behavior when we refine the exoarchitectural description (Fig. 1.6).

For many purposes, these two levels suffice for the complete abstract specification of the hardware system. In some situations, however, a further set of architectural characteristics exists that is, in part, abstracted away at the endoarchitectural level. These collectively define the microarchitecture.

MICROARCHITECTURE. The logical structure and functional capabilities of the hardware system as visible to the microprogrammer.

Note that many aspects of microarchitecture (e.g., the functional units, local store, main store, etc.) will also be part of endoarchitecture; however, the latter need not, in general, include details of the microprogram control unit—details that are essential to the microprogrammer.

Microarchitecture as a separate architectural level becomes especially discernible in systems with writable control stores. Using such systems, the microprogrammer may create new endoarchitectures (and consequently, new exoarchitectures) on a given microarchitecture. Thus, we may view an endoarchitecture as the outcome of implementing a particular microprogram on a given microprogrammable machine (Fig. 1.6).

Example 1.2

In the CAP system represented in Fig. 1.5, the machine's exoarchitecture is defined by the level 2 system. Its characteristics include, in addition to a regular instruction set and a fairly conventional bank of local store registers, special machine instructions for manipulating capabilities, entering and leaving protected procedures, and creating capability segments (i.e., segments consisting only of capabilities). Other aspects of the exoarchitecture include the special data type "capability" and its representation in memory, and a set of capability registers used to hold active capabilities.

CAP's endoarchitecture is represented in Fig. 1.5 by the level 1 system. Among the additional features visible at this level are a "slave" store or cache for reducing the memory access time, registers used as workspace by the microprogram, and the details of the instruction fetch-execute cycle. The latter includes as an important component the capability-based memory access validation phase. Note that although the algorithms performed by the microprogram are part of the endoarchitecture it is not necessary to include the microinstruction formats, their encoding schemes in the microstore, or the microinstruction sequencing logic in order to understand the endoarchitecture. Internal details of the microprogram unit are thus invisible at this level but must of course be available to the microprogrammer. These details are part of the CAP microarchitecture (the level 0 system).

1.3 CONCEPTUAL INTEGRITY OF ARCHITECTURES

The central problem facing the computer architect lies in designing multilevel architectures that must satisfy externally given requirements and constraints on the one hand, and yet exhibit *conceptual integrity* within and between architectural levels on the other. I use this term in Brooks's (1975) and, later, Myers's (1982) sense to imply collectively three related concepts: the notion of architectural unity in the design of the system in question, a cohesiveness among all its parts, and a lack of irregularities in its structure. These objectives are difficult enough when restricted to a single architectural level; they become even more problematic when required to hold true between levels.

Example 1.3

Myers (1982) has cited several features of the IBM System/370 exoarchitecture that weakened its architectural integrity. Among these are the following:

1. The general purpose registers are almost, but not quite general purpose. For instance, the translate and test instruction always uses registers 1 and 2.
2. For any instruction that requires the use of multiple general purpose registers, two consecutively numbered registers must be used. The instruction must name the lower numbered register and this must have an even number.

Example 1.4

The Nanodata QM-1 has two levels of microarchitecture, of which the lower level, termed "nanoarchitecture," is the interface provided to the "nanoprogrammer." When one is writing into "nanostore," using the command WRITE NS, for instance, if an invalid byte address is specified, the addressed word in nanostore is read.

Example 1.5

In his critique of the PDP-11 condition code structure, Russell (1978) has shown several incongruities. First, incrementing (by using the INC instruction) is not functionally equivalent to adding 1 (by using ADD), nor is decrementing the source the same as subtracting 1 (by using SUB). Second, dividing by 2 (by using the arithmetic shift right instruction) may cause overflow. These features arise out of inconsistencies in the setting of condition codes across the PDP-11 instruction set.

Since exoarchitectures and microarchitectures are directly subject to user scrutiny, the lack of conceptual integrity at these levels is more easily detectable than in the endoarchitecture. Sometimes, however, a lack of unity between architectural levels may also become apparent to the user.

Example 1.6

A case in point is the conflict between the need to specify the precise point at which an interrupt occurs and the use of pipelining to enhance processor performance. The former is, at least in part, an exoarchitectural attribute, while pipelining is, in many systems, an attribute of the endoarchitecture that is abstracted away at the higher level. Since a pipelined system operates on several instructions in parallel, possibly out of their original sequence, it may not be possible for the machine to designate the precise instruction that generated the interrupt-provoking error. This leads to the so-called *imprecise interrupt,* a situation in which the hardware may at best indicate to the interrupt-handler that an error or an event of a certain class has occurred in the region of a certain instruction in the original program.

Ultimately, many of the incongruities and anomalies that detract from a computer's architectural integrity originate in possible conflicts of interest between architectural levels, particularly between exoarchitecture and endoarchitecture. First, the responsibility of the former is to provide some desired or prespecified degree of functionality at the acceptable level of performance or cost/performance ratio set for the system; the architect does so through appropriate selection of primitive components and the use of hardware/firmware information processing principles. Sometimes, unfortunately, functional needs and cost/performance objectives clash, and comprises have to be made. Conflicts may also arise between architecture as a whole and extratechnical (e.g., marketing) considerations. Almost invariably it is the functional needs (i.e., exoarchitectural features), that are the first to be sacrificed.

1.4 BIBLIOGRAPHIC REMARKS

An important early contribution to the theory of complexity is Simon (1962). This paper also appears as a chapter in Simon (1981). The hierarchic, multilevel nature of computers is discussed at length by Bell and Newell (1971), and in Dasgupta (1980b). For further discussions of hierarchy theory and complex systems—both natural and artificial—the reader is referred to a collection of articles edited by Pattee (1973), Whyte et al. (1969), and a paper by Parnas (1974).

The basic idea underlying the use of capabilities as a means of memory protection is discussed in Wilkes (1975), while the CAP system is described in some detail by Wilkes and Needham (1980). The important concept of information hiding as an aspect of systems design is due to Parnas (1972).

The CFA project and its objectives, program, and results are presented in a series of papers, notably Fuller, Stone, and Burr (1977) and Fuller, Shaman, et al. (1977), while an interesting application of formal machine descriptions to architectural verification—particularly in the context of the CFA project—is described by Barbacci and Parker (1978).

The main definitions of computer architecture discussed in this chapter are based on Myers (1982), Reddi and Feustal (1976), Dennis et al. (1979), and Baer (1980). Myers (1982) is also a thorough and original treatment of language-directed exoarchitectures, while Baer (1980) offers an excellent, up-to-date text on all aspects of computer architecture. For another viewpoint on the concept of architecture, the reader is referred to Zemanek (1980). The terms "exoarchitecture" and "endoarchitecture" were first introduced by Dasgupta (1981).

The notion of conceptual integrity in the particular context of software design is discussed by Brooks (1975). Its implications for architecture are described by Myers (1982).

The architecture of the QM-1 is detailed in the Nanodata QM-1 Hardware User's Manual (1979); brief discussions of this interesting two-level, user-microprogrammable machine are available in Rosin, Frieder, and Eckhouse (1972), Agrawala and Rauscher (1976), and Salisbury (1976). The discussion on the PDP-11 condition code is based on Russell's (1978) work.

An exhaustive treatment of pipelined computers, including the problem of imprecise interrupts, is given in a recent book by Kogge (1981). This problem is also discussed in a survey paper by Ramamoorthy and Li (1977).

TWO

THE INFORMAL DESIGN PROCESS

2.1 ARCHITECTURE AS CRAFT

In the previous chapter, three distinct, though closely related, architectural levels were identified, each with its own distinct function. Of these, exoarchitecture and endoarchitecture are common to all conventional computers—that is, machines whose users are writers of software—while microarchitecture, though also evident in the conventional computer, is especially discernible in systems with writable control stores—that is, computers whose users are, or include, writers of firmware.

Given such an architectural hierarchy, the main issue of architectural design is one of developing the detailed structure of each level so as to meet the requirements and constraints specified for it, minimizing the mismatch between levels, and maintaining the conceptual integrity of the system as a whole.

It would seem apparent, then, that the computer architect is posed with a rather fundamental problem: the intellectual management of complexity. This is, of course, an insidious problem that faces all concerned with the design and specification of systems. The unusual situation in computer architecture that distinguishes it from other systems design activities (e.g., programming, large-scale engineering, building architecture) is that there exists almost no theoretical or symbolic framework to guide the architectural design process.

To understand this, consider the conventional method of designing architectures. We shall name this the *informal design process*. It is characterized by, among other things, the following attributes:

1. The documentation of the design is usually a combination of informal block diagrams and prose descriptions. The design is, as a result, ill defined with respect to both form and function (or behavior).
2. The dynamic aspects of the architecture, in particular the interaction of hardware/firmware processes, are usually specified incompletely. Even when this is not so, because of the informal method of their description they are liable to be ambiguous and susceptible to misinterpretation.
3. It is extremely difficult, even in principle, for the design to be validated for correctness without constructing and testing the physical system.
4. It is virtually impossible to investigate, manipulate, or alter a design, or evaluate alternative architectures for performance characteristics, without constructing and testing the physical system.
5. The sequence of design decisions and the rationale for them are seldom documented explicitly.

These are all, clearly, serious problems of description and communication. The process of designing architectures thus remains essentially a private experience and enterprise restricted to the original architects themselves. The reader of a design description must frequently guess why a particular decision was made, fill in gaps or make unverifiable assumptions where the design is incompletely specified, and extrapolate and draw unverifiable conclusions where design features are simply omitted.

The last three characteristics are, in addition and still more seriously, manifestations of the primitiveness of computer architecture as a design activity. For, according to design theorists, notably Jones (1970), the lack of a symbolic medium in which to capture the shape of a product and the reasons for the shape, and the consequent inability for one to experiment with the design (in contrast to experimenting with the product itself), is one of the hallmarks of the *craft* stage of design, or what Jones has called "craft-by-evolution."

The stage that follows, or historically has followed, the craft stage, is termed by Jones (1970) "design-by-drawing." The transition from craft-by-evolution to design-by-drawing is marked fundamentally by the separation of thinking from making: the replacement of the product itself by scale drawings as the medium for experiment and change. This transition came about, according to Jones, when the product being designed became too large or complex for a single craftsman to conceive and construct alone. The scale drawing, in effect, was a first, vital, and huge step in the cognitive management of complexity, since it greatly enhanced the craftsman-turned-designer's perceptual span.

It is precisely this inability to provide the architect or reader (i.e., one who wishes to understand the design) with the required perceptual span or, more appropriately, cognitive capacity, that imbues conventional computer architecture with the attributes of a craft rather than a design discipline.

In the rest of this chapter, I shall give examples of two systems that were, so far as this author can tell, designed informally or at best semiformally. The systems discussed have been constructed and are operational, and they are all paradigmatic in respect of the objectives they were intended to meet. However, in each case we shall see that by the very nature of the design process employed, questions concerning correctness, efficiency, or clarity of the architecture design cannot be answered satisfactorily.

Example 2.1

Consider the MU5, described by Morris and Ibbett (1979). Its most important endoarchitectural features are its primary and secondary instruction pipelines. We are interested here in one part of the primary pipeline, namely, the method by which the pipeline system handles the effect of branch instructions.

In order to supply instructions to the execution unit at a rate that matches the latter's execution rate, instructions are fetched from the main store well in advance of the time at which they will be required by the execution section. The instructions thus fetched are also buffered in preparation for execution.

The technique of prefetching and buffering instructions is perfectly satisfactory as long as the instruction sequence is branch-free. It is also conceptually simple; one does not really need, for the purpose of understanding, a formal description. The problem arises when a transfer of control takes place because of a conditional or unconditional jump instruction. Clearly, all the prefetched instructions must be abandoned, and a new sequence of instructions fetched from the "jump to" address. This would result in a drop in the pipeline's performance, since it must wait for the new instruction sequence to be accessed and fetched.

There are two aspects to this design problem worth considering. First, the designer has a need to examine alternative techniques for minimizing performance degradation resulting from the execution of jumps, and to select the one that appears most effective given the operating environment of this particular system. Second, any proposed solution to this design problem will involve at the least a rather elaborate, hardware-implemented control mechanism for the pipeline.

The informal design process is hopelessly inadequate for coping with either of these problems. Since the next chapter discusses in detail the characteristics of what I shall call the *formal design process,* I shall forgo any elaboration at this point. However, it must be noted here that the MU5 designers in fact went halfway toward the goal of formal design: they used simulation to study techniques for anticipating branches and their effects on performance, and based their final design on these studies. The simulated model served as a symbolic representation of the pipelined architecture.

On the other hand, the actual specifications of this pipelined architecture as presented in numerous sources remain wholly and characteristically informal.

Quite apart from the difficulty of comprehending such informal designs, this separation of behavior modeling of architecture from its actual specification (to the extent there is a specification) is both artificial and undesirable. There is no obvious way of establishing the equivalence between the model used in the simulation and the informally specified architecture; nor does there exist a rigorous method of translating a simulated model into a design description, or vice versa.

Example 2.2

The basic idea of capability architecture appears simple. Main store comprises words of two types: data words (for holding both data and instructions) and capabilities. A capability consists of a segment identifier (or information about where it is in memory) and a specification of the access rights afforded to that segment. A capability is thus an extension of the notion of segment descriptors familiar in virtual memory systems.

In order for a process P to access a segment, it is necessary for P to possess a capability for it. Thus the process P will have, in one or more segments called *capability segments,* a set of capabilities that will determine both the segments to which P has access and the precise nature of these access rights.

In such an architecture, the conventional instruction execution sequence is modified according to Fig. 2.1. The central processing unit (CPU) contains a bank of capability registers (similar in concept to a segment table in segmented virtual memory systems) that contains capabilities for all the segments currently residing in main store. When process P is running, during the execution of each of its instructions, the capability registers are checked to determine whether the instruction opcode is consistent with the access rights for the segment s being referenced. If not, a trap will take place and control will be transferred to an appropriate error handling routine.

In such a system, protection depends on the assumption that capabilities cannot be forged. That is, it must not be possible for an arbitrary process to create a capability for a segment and load it into a capability register. This privilege must be available only to that part of the system designed to do this. The purpose of having two types of words, data and capability, is that only those routines or processes having capabilities for capability words

ARCHITECTURE AS CRAFT

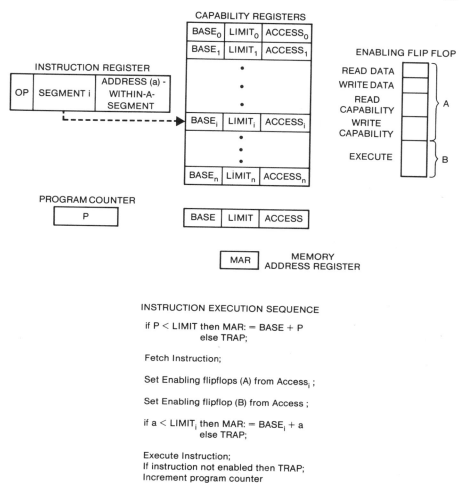

Figure 2.1 Instruction access in a capability-based machine (based on Wilkes (1975)).

(i.e., possessing C-type capabilities) may access and transfer them into capability registers. Possession of a capability for data words (i.e., a D-type capability) allows such words to be accessed and modified.

The routine responsible for creating capabilities must possess both C-type and D-type capabilities for the same segment. It is then possible for the routine to construct a capability and place it in this segment (which will be a capability segment), using its D-type capability for the segment, and then load the capability into a capability register, using its C-type capability.

In the Cambridge CAP computer described by Wilkes and Needham (1980), this protection mechanism has been effected by a combination of hardware, firmware, and software. A capability store consisting of 64 registers is implemented within a hardware capability unit. The firmware consists of microroutines which, in addition to interpreting the CAP's instruction set, is responsible for controlling memory access and enabling the capability unit to validate an access by searching associatively through the capability store. Finally, the CAP operating system is responsible for the management of user processes within the environment provided by the hardware and firmware. It does so by using a set of special instructions specially provided for the manipulation of capabilities.

The key question raised for the CAP or any similar system is, is it possible to demonstrate that this design is correct? Here, the correctness criteria may be formulated as follows:

1. It should not be possible for a process to have access to a segment without having a capability for the segment.
2. Conversely, if a process has a capability for a segment and requires access to it, it must not be barred from doing so.
3. Capabilities may be created by only those processes or routines having explicit authority to do so.

Notice that in such systems, although performance is important, the primary question is one of correctness of the protection mechanism.

Clearly, in order to validate the design with respect to these criteria, one must have a method of specifying:

1. The nature of the operating system functions responsible for protection.
2. The functions of the special set of instructions used by the operating system and the context in which they are used.
3. The tasks performed by the firmware and the contexts in which they are performed by the capability unit.

It is also necessary to understand and specify the policy for the transfer of control between software, firmware, and hardware.

Without an appropriate armory of descriptive tools and concepts, which is wholly absent in the informal design process, it is virtually impossible to respond to the question of correctness for this system.

2.2 BIBLIOGRAPHIC REMARKS

The evolution of design methods is discussed by Jones (1970). A detailed and interesting examination of the various theories proposed over the centuries about the nature of designs and, in particular, how biological and evolutionary analogies have been applied to design, especially in building architecture and the applied arts, is given in Steadman (1979). Other sources of articles on design theory are March (1976), Spillers (1972), Broadbent (1973), Glegg (1973), and Friedman (1980). Finally, the monograph by Alexander (1964) contains many ideas of importance to the design of complex hierarchic systems in general.

The MU5 computer system has been described in many papers. The most detailed and recent treatment is Morris and Ibbett (1979); a short paper by Sumner (1974) provides a retrospective view. My discussion of the CAP system is based on Wilkes and Needham (1980).

THREE

THE FORMAL DESIGN PROCESS

3.1 INTRODUCTION

An architectural design process will be said to be formal if it satisfies the following minimum characteristics:

1. There exists a formal method (or methods) of describing the design.
2. There exist formal principles for validating the design for correctness (against well-defined correctness criteria) without having to construct the physical system.
3. The design can be manipulated and evaluated for its performance characteristics without having to construct the physical system.

The heart of the formal design process, then, is the availability of one or more *formal pictures* of the design—pictures that are objective and formalize the intuitive, imprecise, ambiguous, and subjective conceptualization that is characteristic of informal design. Note that several alternative (but complementary) formal pictures may be brought to bear on a design problem. Thus, for example, Petrie nets may be used to represent a design for the purpose of demonstrating some of its correctness properties, while the same design may be mapped into a simulation language in order to test it for behavior and performance.

The formal design process, then, is centrally concerned with description. However, it will also be observed that as soon as the notion of formal

WHAT IS METHODOLOGY?

description is introduced, the concept and the possibility of a design methodology concomitantly arise. In the informal design process, the absence of formal description almost automatically rules out methodology; the formal design process immediately admits it. Formal description becomes, so to speak, a sine qua non for the development of a methodology of design.

Let us, then, enlarge the definition of the formal design process to include, in addition to the properties already stated, the following:

4. There exists a communicable and teachable methodology of design.

3.2 WHAT IS METHODOLOGY?

It is necessary at this point to clarify the use of the term "methodology" since a cursory glance at the relevant literature reveals that this term, like "structured programming" and "distributed processing" has attained the dubious status of a buzzword with many different (and sometimes vague) connotations.

Methodology, according to the *Concise Oxford Dictionary,* is "the science of method; [the] body of methods used in a particular branch of activity." The *Fontana Dictionary of Modern Thought,* edited by Bullock and Stallybrass (1977), defines the word "in the narrowest sense [as] the study or description of the methods or procedures used in some activity." However, this same source goes on to say, "some scientists use the word as a more impressive-sounding synonym for method."

Thus in its purest sense, the word refers to the study of methods—for example, the methodology of science is the study of method or methods whereby scientific knowledge is acquired. However, as the above quotes indicate, the terms "method" and "methodology" are very frequently used interchangeably, not the least by computer scientists. The *Oxford Dictionary* does provide an alternative definition in which methodology refers to a collection of methods used in some branch of activity, and in our context—that is, in the context of design—this appears to be the appropriate definition.

As Freeman (1980) has pointed out, a design *method* is a way of carrying out the design. More specifically, it involves identifying the decisions to be made, knowing how to make these decisions, and determining the order in which they should be made. Typical and well-known examples of design methods are the topdown, outside-in, inside-out, and bottom-up methods. Note that these methods essentially differ in the way they direct the design development process.

The problem is that in reasonably complex design problems (of the kind commonly encountered in both hardware and software systems design), the selection and application of a single method is usually hopelessly inadequate. Depending on the rigor that the designer wishes to impose on the total design process, the context of the design problem, and the organization of the design team, it may become necessary to bring to bear one or more of the following:

1. A formal view or model of the entire design situation.
2. A set of design methods that suitably complement one another, together with the rules for applying them.
3. A collection of descriptive, documentation and analytical tools.
4. Management techniques for the control and evaluation of the overall design process.

Following Freeman (1980) then, I shall use the term "design methodology" to mean an organized, coherent set of methods, models, description tools, and management techniques that may be applied symbiotically to the systematic and rational development of a design.

Example 3.1

The WELLMADE program design methodology of Boyd and Pizzarello (1978) may be summarized as follows. Given the problem and the programming language to be used:

1. Define an abstract machine M_i.
2. For the program P_i that will run on the abstract machine M_i, identify the data structures to be used by P_i.
3. Develop a functional specification (in the form of input-output assertions) for P_i.
4. Construct a skeleton program for P_i.
5. Refine the program skeleton to obtain P_i.

Now machine M_i must be defined by a program or set of functions running on a lower level machine M_{i-1}.

Thus WELLMADE uses an iterative topdown method in the sense that each iteration consists of:

1. Definition of the abstract machine M_i.
2. Construction of the program P_i that runs on M_i.

Successive iterations deal with the refinements of the abstract machine derived in the previous iteration (Fig. 3.1). In addition to the development method, the methodology incorporates three distinct notational systems—a specification language for representing assertions, a program design language, and a documentation language. It uses such formal constructs as predicate transformers and abstract data types.

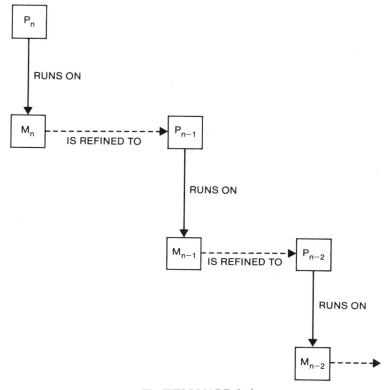

Figure 3.1 The WELLMADE design process.

Example 3.2

Consider the modeling and simulation methodology of Zurcher and Randell (1968). This uses, but distinguishes between, the twin concepts of *hierarchical modeling* and *levels of abstraction*. The former describes the activity of initially representing the system in a suitable algorithmic form and then replacing the algorithm with a sequence of calls on the next set of designed components. The new algorithm thus represents how the function of the original component is obtained in terms of these subcomponents. Continuing in this fashion, the designer describes the behavior of the system by a simulation program organized as a hierarchy of procedures.

A system may exhibit several levels of abstraction. Each level would consist of a simulation program constructed of a hierarchy of procedures as just described.

The overall modeling procedure may then be summarized as follows. At first, a single level simulation program is constructed in hierarchical fashion and used to model and validate the initial design. A second, lower level of abstraction is then introduced. This incorporates features that were considered inappropriate in the initial model. The second level thus contains variables that appropriately represent the system at this abstraction level. Their relationship to the original set of variables must also be defined.

The program at this second level of abstraction will also be developed hierarchically. In this way, a set of levels will be constructed that models the system at several meaningful abstraction levels, until the design is expressed at the final level of detail required.

The Zurcher-Randell (1968) methodology was considerably influenced by the outside-in method proposed by Parnas and Darringer (1967) in their SODAS methodology. Although the methodology as described does not presuppose any particular language, it does necessitate that descriptive tools be available for specifying the different abstraction levels.

Example 3.3

A methodology developed by Estrin and co-workers, described in Estrin (1978) supports a multilevel design procedure for software and hardware development. The methodology is based upon, and is supported by, an interactive computer-aided system called SARA, which provides an integrated collection of languages and tools for modeling and design.

SARA supports both bottom-up composition and top-down partitioning procedures; for any given design problem both these methods would generally apply since beginning with the initial set of requirements and evaluation criteria, the designer constructs a behavioral model of the system and tests this model against the available set of predefined building blocks. If the model is found to be too high-level relative to the available building blocks, it is decomposed into more elementary components. Top-down partitioning would continue in this fashion. Eventually one or more components would emerge that could be composed from available building blocks. Bottom-up composition would then apply to each such component so that a composite model consisting of predefined building blocks would be obtained, tested, and evaluated.

The theoretical underpinnings of this methodology are provided by the UCLA *graph-model of computation*. The tools for description, modeling, and simulation include a high-level language called SL1 that lets a designer specify the (possibly multilevel) structure of a system using a small set of primitives, and a behavior modeling tool called GMB with which the designer can create graph models of the flow of control and data through the system.

3.3 THE LIMITS OF FORMAL DESIGN

It is, of course, quite unlikely that a formal design process could completely replace informal design. To avoid any misunderstanding on this point, let me emphasize that no such claim is being made here. For there is at least one factor that will set the limits on the scope of formal design.

When a problem is presented to the computer architect, he or she draws upon knowledge of established architectural principles, familiarity with previous designs, and experience of similar design projects in the past; on the basis of these the architect may develop an initial (perhaps tentative) concept of the overall architectural form. The basic, dominant architectural style would probably emerge at this stage (the question of style in architecture is discussed at greater length in Chapters 12 and 14). I shall continue to use the term "informal design" to describe this phase, since it is usually an integral part of the informal design process identified in the last chapter.

The formal design process, clearly, cannot substitute for this part of the informal design. It will, however, lend a sharp and precise form to the initial set of concepts and will, further, provide the basis for a detailed development of the architecture and its rigorous evaluation and validation.

Thus I suggest that informal design (in the sense described here) and formal design will play complementary roles in the overall design process. In this light, the interesting question is the extent to which we can expand the scope of the formal design process—that is, the extent to which we may formalize the design of computer architecture.

3.4 BIBLIOGRAPHIC REMARKS

The definitions of the term "methodology" presented in Section 3.2 are from the sixth edition of the *Concise Oxford Dictionary* edited by Sykes (1976) and the *Fontana Dictionary of Modern Thought,* edited by Bullock and Stallybrass (1977). For a compendium of writings on the methodology of science the reader is referred to Suppe (1977). My characterization of the term "design methodology" was largely influenced by Freeman (1980), which is also an interesting discussion of the general nature of design.

The WELLMADE methodology is described in Boyd and Pizzarello (1978) while Example 3.2 is based on Zurcher and Randell (1968). SARA has been described in several papers, notably Estrin (1978), and Campos and Estrin (1978).

FOUR

ISSUES IN LANGUAGE DESIGN

4.1 INTRODUCTION

It was suggested in Chapter 3 that in order for a formal design process to take place, a system of notation must be selected for the precise specification of the design. In the case of complex, multilevel, hierarchic systems, the design process may involve a succession of stages, each of which represents the system at a particular level of abstraction; in such situations, a system of languages or notations may be used.

This chapter is concerned with some of the general issues encountered in the design of languages that may be used for the design and specification of computer architectures. I shall call these simply *architectural languages;* they form a subclass of the class of languages commonly referred to as *computer hardware description languages* (CHDLs).

It must be pointed out that the idea of CHDLs is not particularly recent. Dietmeyer and Duley (1975), for instance, have traced their origins back to the work of I. S. Reed in the 1950s. However, it was really in the mid-1960s that the first systematic design efforts in CHDLs were reported—perhaps largely inspired by the successful developments of the early high-level programming languages.

This chapter is based on a section from the article "Computer Design and Description Languages" by S. Dasgupta, appearing in *Advances in Computers,* Vol. 21 (Ed.) M. C. Yovits, Academic Press, N.Y., 1982. Adapted with permission of the publisher.

LEVELS OF ABSTRACTION

The number of language proposals, both for hardware and firmware specifications, has since increased steadily. Yet, as noted by many observers, the fact remains that unlike the situation in the software domain, CHDLs have not succeeded in entering the computer designer's common culture. This is especially true in the case of architecture.

In assessing the many existing languages for hardware descriptions, and in reflecting on the requirements of a formal design process for architectures, a number of significant dimensions emerge that collectively form a *description space*. A given architectural language (or, more generally, a CHDL) may then be viewed as a point (or more realistically, a localized region) within this space, according to its specific characteristics. The main dimensions that we may identify are:

1. Levels of abstraction that a language describes.
2. Whether the language is operational or functional.
3. Whether the language is procedural or nonprocedural.
4. The ability of the language to specify behavior and/or structure.
5. The influences of programming languages.

In the sections that follow, each of these factors is discussed in some detail.

4.2 LEVELS OF ABSTRACTION

The main point of Chapter 1 was that a typical computer system is hierarchically structured and that even its hardware component can be decomposed into several distinct and meaningful levels of abstraction.

In designing an architectural (or more generally, a hardware description) language, the choice of its level of abstraction is one of the significant factors the designer must consider. Thus, a given language may be characterized by whether it describes the machine at one or more of its architectural levels.

It is also important to distinguish between architectural and *register-transfer* languages. To take some examples, DDL, developed by Dietmeyer and Duley (1975) and CDL, developed by Chu (1972) are archetypal register-transfer languages in that they describe computer structures in terms of such primitive types as terminals, registers, delays, clocks, memories, and combinational circuits. The primitive types bear obvious correspondences with common MSI building blocks.

A language called ADL, developed at the University of Manchester and described by Burston, Kinniment, and Kahn (1978), is more difficult to classify.[1] The elemental components in an ADL description are basic sub-

---

[1]This language, the name of which is an acronym for Asynchronous Design Language, is not to be confused with Leung's (1979) ADL (Architecture Description Language) which was developed at the Massachusetts Institute of Technology primarily for the description of packet communication systems.

blocks with defined input/output data and control ports, primitive actions (e.g., for purposes of data transfer, synchronization, or control transfer), and connections (which describe interconnections or subblocks). Inasmuch as within the language definition these entities are not related to specific classes of integrated circuits, ADL may be regarded as an endoarchitectural language. However, since the construct for declaring basic subblocks is mostly useful for specifying registers, shift registers, counters, and the like, ADL also exhibits characteristics of register-transfer languages.

Example 4.1

This hybrid aspect of ADL may be illustrated by Fig. 4.1. This defines a module or block with data input ports INST_FROM_SAC (128 bits wide) and LIS_FROM_STORE_REQ (one bit wide), and control input ports DATASENT and LIS. The block IBU is itself composed of the (simpler) subblocks INPUT_BUFFER and INPUT_MECH. The former is a basic subblock of type REGARR816. Thus, the description assumes the existence and definition of REGARR816 (which denotes an array of 8-bit registers) in a library of basic subblock types.

The subblock INPUT_MECH is not basic; it is defined by the system designer. As with ADL blocks in general, the inputs to and outputs from this subblock are contained in its declaration. In addition, the permanent interconnection of blocks and subblocks is given by the "CONNECTION" declaration. Thus, for example, the input INST_FROM_SAC to the outermost block IBU is permanently connected to the input port IIN of the subblock INPUT_MECH.

The last component of INPUT_MECH is the nonterminating procedure 'BEGIN TL_INPUT :...'END', the execution of which dictates the flow of data and control signals through the block.

The important point to note about this example is that, in general, the underlying hardware is implicit. For instance, it is not evident from the description how INPUT_MECH is to be realized in hardware. This deliberate hiding of specific hardware interpretations with a concomitant emphasis on the logical structure and behavior is, I suggest, the basic characteristic of the architectural level of abstraction.

At the same time, note that ADL permits reference to predesigned modules with specific hardware interpretations—for example, the use of a particular register type REGARR816. The visibility of such specific types of circuit entities is the main characteristic of the register-transfer level. For instance, a description in DDL—a typical register-transfer language—is a specification of hardware in terms of register, terminals, clocks, and automata. Similar entities exist in Chu's CDL (1972) and AHPL, an APL-like language developed by Hill and Peterson (1978).

```
BLOCK
    IBU['INPUT' INST_FROM_SAC[0:127], LIS_FROM_STORE_REQ
        'CONTROL IN' DATASENT, LIS
        ....]

BASIC SUBBLOCK REGARR816
    INPUT_BUFFER['INPUT' IIN[0:127], IADDRESS[0:2]
                 'OUTPUT' IOUT[0:15]
                 'CONTROL IN' ILOAD, ISELECT
                 'CONTROL OUT' ILOADDN, ISELECTDN]
....
SUBBLOCK
    INPUT_MECH['CONTROL IN' IBUFFER_EMPTY, DATASENT1, ILOADDN1
               'CONTROL OUT' LOAD_INPUT_BUFFER, IBUFFER_FULL1]
    'CONNECTION' IIN <- INST_FROM_SAC
    'CONNECTION' 'CONTROL'
        DATASENT1 <- DATASENT, ILOAD <-LOAD_INPUT_BUFFER,
        ILOADDN1 <- ILOADDN

    'BEGIN'
        T1_INPUT:'WAIT FOR' IBUFFER_EMPTY
                 'WAIT FOR' DATASENT1
                 'SET' LOAD_INPUT_BUFFER
                 'WAIT FOR' ILOADDN1
                 'SET' IBUFFER_FULL1
                 'GOTO' T1_INPUT
    'END'
....
```

Figure 4.1 An ADL description (copyright © 1982, Academic Press; reprinted with permission).

4.3 THE OPERATIONAL-FUNCTIONAL DICHOTOMY

An issue of particular relevance in the design of an architectural language is the choice between an *operational* description, in which the behavior of the system is defined in terms of an abstract program, and a *functional* description, which prescribes behavior according to some logical relationship between the arguments and results of operators.

This problem is, of course, not peculiar to hardware or architecture; it has been debated extensively in the context of software specification languages, specifications of abstract data types, and programming language semantics. In fact, the operational/functional dichotomy becomes relevant whenever the behavior of some computational entity is the issue at hand.

We may show the contrast between the two approaches with an example. Consider the data type STACK. A functional specification of this type will typically be along the lines shown in Fig. 4.2. Here, the first line, in addition to giving the name of the stack, provides two input parameters associated with the type. The parameter ELEM_TYPE is a variable that may take on as its value the name of any defined type. The parameter N is an integer variable restricted to positive values.

The section on syntax lists a set of six operations, the last of which is specially marked to indicate that it is an auxiliary operation, required to define the semantics but hidden from external procedures that use the STACK datatype. The semantics of these operations are then given as axioms in the section on semantics. For example, the first axiom states that the sequence of PUSH followed by POP leaves the stack in its original state. Similarly, the second axiom defines the result of TOP in terms of the most recent PUSH operation. The remaining axioms are equally straightforward.

The section on restrictions provides a precondition and two exceptions. The precondition states that one cannot use POP on an empty stack. The second restriction indicates that an attempt to perform a TOP operation on a newly defined stack will lead to failure. The last restriction states that any failure that arises from an attempt to use PUSH to add an element onto the stack implies that the stack is already full.

This functional specification may be contrasted with the more common operational definition shown as Fig. 4.3. Here, the data type STACK has been defined in terms of three global variables—stk, stkptr, and limit—and the effects of the five operations shown on these variables. Further, their actions are described in terms of abstract programs written in an abstract programming language.

As Melliar-Smith (1979) has noted, the problem with the operational approach lies in the possibility of overspecification, since it defines not only what the operations do but also how to compute them. Aspects of what are normally considered as implementation details enter into the description. For example, the specification of Fig. 4.3 binds the data type STACK to a particular data structure, namely an array. Notice, in contrast, that Fig. 4.2

```
Type STACK[ELEM_TYPE : TYPE; N : INTEGER where N > 0];
```

Syntax

```
    NEWSTACK : -> STACK
    PUSH : STACK X ELEM_TYPE -> STACK
    POP : STACK -> STACK
    TOP : STACK -> ELEM_TYPE
    ISNEW : STACK -> BOOLEAN
   *DEPTH : STACK -> INTEGER
```

Semantics

```
    STK : STACK, ELM : ELEM_TYPE
    POP (PUSH(STK, ELEM)) = STK
    TOP (PUSH(STK, ELEM)) = ELM
    ISNEW (NEWSTACK) = TRUE
    ISNEW (PUSH(STK, ELM)) = FALSE
    DEPTH (NEWSTACK) = 0
    DEPTH (PUSH(STK, ELM)) = 1 + DEPTH(STK)
```

Restrictions

```
    pre (POP, STK) = ~ISNEW(STK)
    ISNEW (STK) => failure (TOP, STK)
    failure (PUSH, STK, ELM) => DEPTH (STK) ≥ N
```

Figure 4.2 Functional specification of the data type STACK (copyright © 1979, IEEE; reprinted with permission).

says nothing about what form STACK should take, thus extending to the implementer a range of possibilities in actual data structures.

Nonetheless, operational descriptions have compelling advantages, not the least of which is that the form of such descriptions is familiar and the descriptions may be readily understood. Functional specifications—especially in their most abstract form—will almost certainly not be accepted as widely as their operational counterparts. However, in certain domains the functional descriptions are becoming increasingly commonplace; for example, in the use of axioms and proof rules for defining the semantics of programming languages. In the case of more complex entities such as operating systems, forms of functional specifications that are relatively less aus-

```
type STACK;
  stk : array[limit] of integer;
  stkptr, limit : integer;
  initially stkptr = 0, limit = 0;

  operation PUSH (inputs item : integer);
     if limit = 0 then ERROR 'STACK NOT CREATED'
        else if stkptr = limit then ERROR 'STACK FULL'
           elsestkptr := stkptr+1;
             stk[stkptr] := item
           fi
     fi
  endop;

  operation POP (outputs item : integer);
     if limit = 0 then ERROR 'STACK NOT CREATED'
        else if stkptr = 0 then ERROR 'STACK EMPTY'
           elseitem := stk[stkptr];
             stkptr := stkptr-1
           fi
     fi
  endop;

  operation TOP (outputs item : integer);
     if limit = 0 then ERROR 'STACK NOT CREATED'
        else if stkptr = 0  u>then ERROR 'STACK EMPTY'
           else item := stk[stkptr]
           fi
     fi
  endop;

  operation NEWSTACK (inputs initlimit : pos integer);
     limit := initlimit
  endop;

  operation ISNEW (outputs new : Boolean);
     if limit > 0 then new := true else new := false fi
  endop

endtype
```

Figure 4.3 Operational description of the data type STACK (copyright © 1982, Academic Press; reprinted with permission).

tere have been or are being proposed, for example the Parnas module (Parnas 1972) and its extensions.

Hardware descriptions have more or less been dominated by the operational approach. A notable exception is a recent proposal by Frankel and Smoliar (1979) that applies Guttag's (1977) algebraic approach to the functional specification of computer architectures.

4.4 PROCEDURAL AND NONPROCEDURAL DESCRIPTIONS

Most of the commonly used programming languages are procedural. That is, there exists an implicit ordering in the execution of statements and this ordering is imposed by the textual structure of the program. Associated with this structure is an implicit instruction counter that, in the course of execution of the program, is automatically incremented except when a branch causes a change in the flow of control.

Procedural programming languages are a natural consequence of the so-called von Neumann model of computer architecture. Along with the model itself, they have remained the dominating paradigm in the design of programming languages. There are, however, several areas of computing systems where computation proceeds in a nonprocedural fashion. By this I mean that if we were to describe such computations in some language, the order of the statements would not necessarily be significant for the order of their execution. Thus, there would be no instruction counter to be automatically incremented or modified as the computation proceeded. Instead, an action within a nonprocedural computation would take place when and only when some particular condition was satisfied. The general form for such actions may be depicted as:

$$
\begin{array}{lll}
\text{cond } 0 & : \text{ actions } 0 & ; \\
\text{cond } 1 & : \text{ actions } 1 & ; \\
\quad \vdots & \quad \vdots & \\
\text{cond } i & : \text{ actions } i & ; \\
\quad \vdots & \quad \vdots & \\
\text{cond } n & : \text{ actions } n & ;
\end{array}
$$

If at any stage of computation condition i is satisfied, then actions i (a set of one or more concurrent actions) will take place. Their outcome may possibly make condition j true, in which case actions j would next take place, and so on. The only situation in which the textual ordering of the statements may have any significance is when at any given time two or more of the conditions hold true. In this case the highest placed set of actions may be selected.

Nonprocedural computations and systems are encountered in a variety of forms. In artificial intelligence (AI), for example, a particular model of information processing takes the form of production systems that essentially describe nonprocedural programs. In the past few years, a novel class of computers, called data flow machines, has been proposed, where an operational unit is activated whenever all the input values necessary for the operation are available on the input lines to the unit (Fig. 4.4). In the context of hardware, nonprocedural computations are of particular interest, since digital systems can be and have been widely modeled and designed as (nonprocedural) state machines. That is, if the system is in state S_i a certain set of actions will take place and the system will undergo a transition to state S_j (say). State S_j triggers another specific set of actions, accompanied by a further change of state. Thus, there exist many CHDLs that are non-

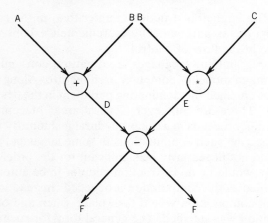

Figure 4.4 Data flow computation.

procedural in style, most prominent of these being DDL, CASSANDRE (Anceau et al. [1971]), and CDL. Appropriately enough, these are all register-transfer languages, since the nonprocedural nature of hardware systems is most prominent at this level.

It must be noted that a language may allow both modes of description. For example, in SLIDE, an input/output (I/O) description language designed by Parker and Wallace (1981), the basic executional entity is the process construct. Although the statements within a process are executed in procedural fashion, as defined by the implied flow of control, the process as a whole is activated nonprocedurally. The condition under which the process becomes active is stated by initialization statements. For example:

> INIT X WHEN condition
> . . .
> PROCESS X ;
>   BEGIN . . .END

A similar capability is provided in the architecture description language S*A designed by Dasgupta (1981, 1982): Individual S*A mechanisms can be prescribed for activation under conditions stated within initialization statements. Examples of this will be given in later chapters.

4.5 STRUCTURE AND BEHAVIOR

A further dimension of the computer description space relates to the distinction between structure and behavior. Thus, a given language may be characterized by whether it allows description of system behavior only, system structure only, or both.

A hardware system has obvious structural characteristics at all levels of abstraction. In very general terms, this structure consists of a collection of entities (the nature of which depends on the level of abstraction) interconnected to form a network.

The pleasant thing about network-like structures is that they can be easily diagrammed. For this reason, the description of hardware systems almost invariably includes, and is sometimes dominated by, structural diagrams. Yet, in spite of their obvious advantage—the immediacy of their visual impact—they suffer from serious disadvantages. Of these perhaps the most important is that network-like diagrams tell one almost nothing about system dynamics—how the system behaves over time and how, in fact, information and control actually flow through the network. Further, except for logic circuits that describe the switching circuit level, or Petrie nets that describe abstract concurrent systems, structural diagrams are rarely formal or unambiguous. Finally, one must note that such diagrams are not readily machine-readable; hence they cannot normally be input to design automation or simulation systems.

Clearly, the nearer the description level is to the level of implementation the greater the necessity for a structural description. Because of this many register-transfer languages provide some capability for the specification of structure. In essence this includes a construct for designing the connection of two or more entities and the means for declaring the structure of the individual entities. Given such a structural description it is usually possible to construct a corresponding structural diagram.

A simple illustration is the ADL description in Fig. 4.1. Here, we have defined an entity called IBU with its input data and control ports. The block IBU is composed of—among other things—REGARR816 (a register array) and a functional unit named INPUT_MECH. Both these have their own input and output ports.

The structure is given final shape by means of CONNECTION statements. Pictorially, the structure will be as shown in Fig. 4.5.

As another example at a lower level, Fig. 4.6 shows the specification of a three-input, one-output multiplexer in base CONLAN, one of the constituent languages of the CONLAN project currently under development (Piloty et al. [1980a]). This circuit consists of four nand gates that are all instances of a single type, nandgate1. (In this particular example, nandgate1 is also defined internally, within the inner DESCRIPTION..END clause, much like an internal procedure; it could as well be defined externally to the multiplexer and be simply referenced here.)

The symbol Δ denotes the delay operator, while the symbol .= denotes terminal connection. The reserved word btm1 denotes the type "Boolean terminal with default value 1." The connection between the terminals is specified in the lower part of the description. The corresponding diagram is given in Fig. 4.7.

In both the above examples, structural and behavioral specifications are

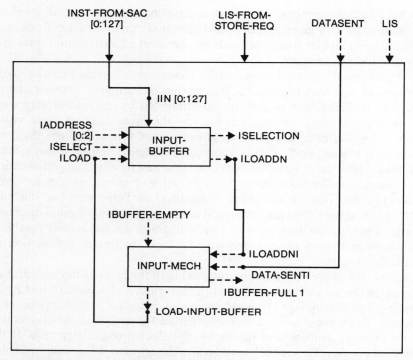

Figure 4.5 Structural diagram for the ADL description of Fig. 4.1 (copyright © 1982, Academic Press; reprinted with permission).

```
DESCRIPTION multiplex (IN a, b, c : btm1, OUT d : btm1) BODY
    DESCRIPTION nandgate1 (IN x, y : btm1, OUT Z : btm1)
        BODY Z .= (x ⌐∧ y) △ 1 ENDnandgate1
    DECLARE i, j, k, m : btm1 ENDDECLARE
    USE g1, g2, g3, g4 : nandgate1 ENDUSE

    g1.x .= a,   g1.y .= a,    i .= g1.z,
    g2.x .= b,   g2.y .= a,    j .= g2.z,
    g3.x .= i,   g3.y .= c,    k .= g3.z,
    g4.x .= j,   g4.y .= k,    d .= g4.z

ENDmultiplex
```

Figure 4.6 Specification of a multiplexer in Base CONLAN (copyright © 1982, AFIPS Press; reprinted with permission).

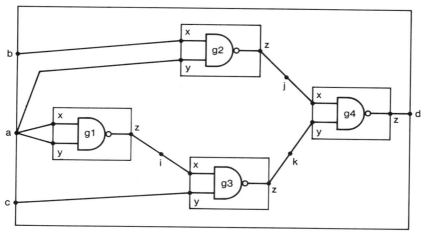

Figure 4.7 Structure of multiplexer (copyright © 1979, AFIPS Press; reprinted with permission).

integrated into a single composite description. This is in general a characteristic capability of register-transfer languages, since it would seem that a structural description by itself would be of limited use. The notable exception to this is when the gross structure of a computer system, a specific system configuration, or a particularly complex or novel structure must be depicted. The level of abstraction here is so high that behavioral

Memory (M):	A component that holds information. The type of memory may be explicitly given as part of the abbreviated notation, e.g., M_p denotes primary (main) memory.
Link (L):	Transfers information from one place to another.
Transducer (T):	A component that changes the presentation of information (e.g., from a nondigital to a digital form), without changing its meaning. A particular type of T may be explicitly indicated in the notation, for example, T. console.
Data Operation (D):	Performs data operations, for example arithmetic or logic.
Control (K):	Activates and controls other components in the system. The type of K may be stated in parentheses, for example, $K(M_p)$ for primary memory control.
Switch (S):	Makes and breaks links between components.
Processor (P):	A composite component consisting of sets of Ds, Ms, Ts, Ss, and Ls together with a control K that is capable of accessing instructions from memory and interpreting them. The type of P may be explicitly indicated; for example, P_c denotes a central processor, P_{io} denotes an I/O processor (channel).

Figure 4.8 PMS notation: primitives (copyright © 1982, Academic Press; reprinted with permission).

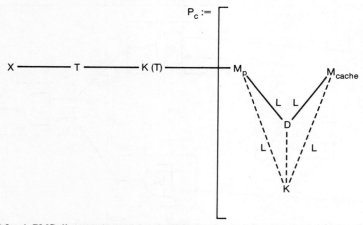

Figure 4.9 A PMS diagram (copyright © 1982, Academic Press; reprinted with permission).

specifications would not be particularly relevant; it is the system structure that is of interest.

The Processor-Memory-Switch (PMS) notation was developed by Bell and Newell (1971) for precisely this purpose. This is a general purpose notation for the description of the structure of information processing systems; it consists essentially of seven primitive component types, each with an associated symbol. These components are listed and explained in Fig. 4.8. A typical PMS diagram, indicating the structure of a simple computer at the architectural level, is given in Fig. 4.9. Here, the solid lines denote data links (paths), and the dashed lines, control paths. The term X denotes the external environment. Although not shown here, additional attributes of the individual components (such as the word size and number of words in M_p) may be shown alongside the components' names.

As a final example of structural descriptions let us consider the SARA system developed by Estrin and coworkers at UCLA (Estrin [1978]). As mentioned in Chapter 3, this is an integrated set of software tools designed to support the systematic synthesis of hardware and software systems. One of SARA's many components is a module called STRUCTURES that allows the designer to describe system structure interactively using three primitives: *modules, sockets,* and *interconnections*.

A module represents an entity whose internal structure is not visible to its environment, and whose only possible communication with the environment is through one or more sockets. Sockets in turn are connected with one another through structures called interconnections. Modules can be nested to form multilevel structures.

Figure 4.10 shows, in SARA's graphic notation, the structure of the entity described (in ADL) in Fig. 4.1. The outermost module is INST_BUF-FER_UNIT with sockets @IPU1, @IPU2, @IPU3, and @IPU4. Nested

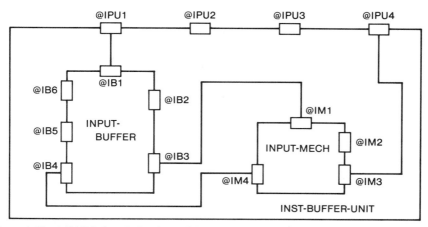

Figure 4.10 A SARA description (copyright © 1982, Academic Press; reprinted with permission).

inside are two modules INPUT_BUFFER and INPUT_MECH, each with its own set of sockets. Some of the interconnections of sockets are shown as lines. SARA includes a language processor called SL1 that can be used to describe this same structure in machine-readable form.

At the level of exoarchitecture, structure is much less important; indeed, it has hardly any significance whatsoever. While PMS-like notations may be applied to the specification of endoarchitectures, the structural relationship of components (especially in the case of von Neumann-type designs) contributes very little to the system's complexity. Endoarchitectural complexity arises primarily from the manifold interaction of component behavior. Thus the predominant responsibility of formal description of architectures is the controlled specification of architectural behavior; the provision, that is, of a means of comprehending and mastering behavioral complexity. For this reason architectural description languages such as ISPS (Barbacci et al. [1978]) and S*A (Dasgupta [1981]) describe behavior rather than structure. Leung's (1979) ADL, on the other hand, allows for both structural and behavioral descriptions. Since ADL is intended for the description of data-flow computers whose structures are likely to be far more complex than those of von Neumann-style machines, structural descriptions seem necessary.

4.6 THE INFLUENCE OF PROGRAMMING LANGUAGES

An interesting dimension of the description space is the extent to which programming languages have influenced CHDL design and the nature of this influence. It seems quite natural for programming languages to provide the lead in this respect: it is, after all, the business of their designers to search for and implement new, effective, and general computational mechanisms. The more general the mechanism the greater its potential for adaptation to specific contexts, including hardware.

The influence of programming languages on the earlier CHDLs is primarily seen in the design of constructs for execution and control. This was particularly the case with high-level microprogramming languages (HLMLs), a special class of CHDL's. For instance, Husson (1970), in his pioneering text on microprogramming, described an early design of a HLML that was strongly modeled on FORTRAN. Eckhouse (1971), in proposing another microprogramming language called MPL, regarded it as essentially a dialect of PL/1.

Among the more conventional CHDLs, AHPL, developed in the early 1970s by Hill and Peterson (1978), is an adaptation and extension of APL. However, though Chu (1965) originally described CDL as an "Algol-like" design language, this influence was probably more in spirit than in content, since there are very few similarities between Algol 60 and CDL.

Since the latter half of the 1970s, the interest in programming language design has shifted to such topics as the identification of powerful data structuring and data abstraction facilities, interprocess communication and synchronizing mechanisms, and the search for reliable software structures. The development of more recent CHDLs reflects this shift in interest. Thus, Leung's (1979) ADL, Dasgupta's S*A (1981) and S* (1978) (the latter is a microprogramming language), and Patterson's (1976) microprogramming language STRUM were all heavily influenced in their data and control structuring capabilities by Pascal. Leung's ADL and S*A include synchronizing mechanisms; S*A, ADL, and the I/O description language SLIDE by Parker and Wallace (1981) contain constructs for modularizing descriptions; and ADL incorporates the monitor construct. The mechanism construct in S*A was influenced by both the monitor and the process concepts present in Concurrent Pascal.

A different kind of software influence is seen in the microprogramming languages EMPL, proposed by DeWitt (1976b), and MARBLE, developed by Davidson and Shriver (1980), and in the CONLAN project reported by Piloty et al. (1980a), all of which rely on the idea of language extensibility. Finally, one must not fail to mention Frankel and Smoliar's (1979) proposal of a register-transfer language based on the work of Guttag and others on algebraic specification systems.

A rather notable exception in this respect is ISPS, which does not appear to have been strongly biased in any obvious way by prior software languages or concepts.

4.7 BIBLIOGRAPHIC NOTES

For discussion of the earlier register-transfer languages, the reader is referred to the survey articles by Barbacci (1975) and Dietmeyer and Duley

(1975). An extensive treatment of architectural languages is given in Dasgupta (1982). The present chapter is based on a section from this paper.

Dietmeyer and Duley (1975) also contains a detailed description of DDL and its applications. The language CDL was first presented in Chu (1965) and later in Chu (1972). The language ADL, developed in Manchester, England, has been described in Burston, Kinniment, and Kahn (1978). The language AHPL is described in detail in [Hill and Peterson 1978], and CASSANDRE, in Anceau et al. (1971).

The problems of system specification, in particular the distinction between, and the relative advantages of, functional and operational specifications, are further discussed by Melliar-Smith (1979). Parnas modules were originally presented in Parnas (1972). A recent extension was developed by Levitt and Robinson (1977). Melliar-Smith (1979) also discusses this topic.

The concept of abstract data types based on a data algebra is due to Guttag (1977). Its application to the functional description of architectures was suggested by Frankel and Smoliar (1979). The development of dataflow computers is currently an active area of research that spans the fields of both architecture and programming languages. Recent discussions may be found in Dennis et al. (1979) and Treleaven, Brownbridge, and Hopkins (1982). Leung's ADL, designed specifically for the description of such machines, was originally discussed in Leung (1979), while a more recent paper, Leung (1981), describes the application of the language to the top-down design of machines.

Discussions of S*A were originally presented in Dasgupta (1981), Dasgupta and Olafsson (1982), and Dasgupta (1982). Later chapters of this book discuss this language in much greater detail. A description of SLIDE may be found in Parker and Wallace (1981), while ISPS is discussed in several sources, notably Barbacci et al. (1978) and Barbacci (1981). The CONLAN effort has been presented by Piloty et al. in a series of papers (1980a, 1980b, 1980c).

A detailed review of the issues in the design and application of high-level microprogramming languages is given in Dasgupta (1980b). Other general surveys include Sint (1980) and Davidson (1982). For discussions of specific proposals the reader is referred to Patterson (1976), Dasgupta (1980a, b), DeWitt (1976a, b), Davidson and Shriver (1980), Davidson (1980), and Klassen and Dasgupta (1981).

FIVE

A LANGUAGE FOR DESCRIBING COMPUTER ARCHITECTURES

5.1 TOWARD A FAMILY OF DESIGN LANGUAGES

We saw in Chapter 1 that computer architecture as a discipline encompasses several (logical) levels of abstraction of the physical machine. The design process for such a system thus involves a succession of stages, each of which represents the system at a particular level of abstraction. In such situations, several languages may be required, each suited to a particular level or set of levels.

Given this possibility it seems quite natural to inquire whether one could develop a family of languages that could be applied to the design and implementation of computer architectures. By constructing languages that are relatively close in their syntax, semantics, and modeling capabilities (in which case we may say that they are *kin* languages), it may be expected that, while on the one hand they may be used independently and separately, on the other, they may also be used symbiotically, so that the transformation from one level to the next can be made relatively painless and within the designer's intellectual grasp. The transformation may even be (partially) automated. Further, all stages of the design may be documented within a unified descriptive framework.

The development of the S* family of languages (denoted symbolically as [S*]) is based on this idea. As envisioned here, [S*] is an open-ended family, in that new members can be freely admitted as required to meet the needs of new architectural levels and paradigms.

Presently, [S*] consists of two completely defined members: S*A, a general purpose procedural language for the formal, operational specification of exoarchitectures and endoarchitectures, and S*, a high-level microprogramming language schema (from which the family derives its name). The latter, in turn, can be *instantiated*, or realized, as a subfamily of machine-specific microprogramming languages. One such fully instantiated language is S* (QM-1), oriented toward a user-microprogrammable computer, the Nanodata QM-1.

After a brief discussion of the features of [S*] as a whole, the remainder of this chapter is devoted to the S*A architectural language. The microprogramming language schema S* and its instantiated version S* (QM-1) are discussed in later chapters. Most of the remaining chapters in this book are devoted to examples of the application of [S*] to the formal, symbolic design of architectures.

5.2 AN OVERVIEW OF THE S* FAMILY

The main characteristics of S*A and S* are compared and summarized in the first three columns of Table 5.1.

Table 5.1 Principle Features of the S* Family

Characteristics	S*A	S*	S*(QM-1)
Primitive data types	bit	bit	bit
Structured data types	seq, array, tuple stack, assoc array	seq, array, tuple stack, assoc array	seq, array, tuple
Synchronizing primitives	yes	yes	no
Constant declaratives	yes	yes	yes
Pseudovariables	no	yes	yes
Channels	yes	no	no
Basic executional statement	assignment statement	assignment statement schema	machine specific assignment statement
Control statements	if..fi, while..do repeat..until, case, call, act retn, exit, goto	if..fi, while..do, repeat..until, case, call, retn, goto	if..fi, while..do, repeat..until, call, act, case, retn, goto
Parallel statements	yes	yes	yes
Machine-timing related constructs	none	cocycle..coend, stcycle..stend, region..regend	cocycle..coend, stcycle..stend, region..regend
Modularization concepts	system, mechanism, procedure	program, procedure	program, procedure, macro

S* is termed a microprogramming language *schema* because its syntax and semantics are only partially defined. For a given microarchitecture M1, a particular language S*(M1) is obtained by completing the specification of S* on the basis of the idiosyncratic properties of M1. The schema S* is then said to be *instantiated* into a particular language S*(M1).

The defined entities in S* include a set of primitive and structured data types, constructs for the description of microprogrammable data objects, and a set of composite control statements, including statements for the specification of parallel action. The primitive assignment statement is only partially defined in S*, since its exact form and meaning may vary considerably from one microarchitecture to another. Thus, given two instantiated languages S*(M1) and S*(M2) for two microprogrammable machines M1 and M2 respectively, the only substantial distinction will be at the level of the primitive statements. I have pointed out elsewhere that for various reasons, the idea of such schemata represents a *satisficing* solution (to use a word created by H. A. Simon [1981]) to the problem of machine independence in high-level microprogramming languages.

From Table 5.1 we can identify the common and distinctive characteristics of the two languages as follows:

1. S*A and S* contain almost the same set of data types, data declaration facilities, and execution and control statements.
2. S* includes constructs that may be used for the specification of the clock-related operations that are so critical in the microprogramming domain. Since such low-level timing issues are conventionally suppressed or hidden at the endoarchitectural or exoarchitectural levels, these constructs are absent in S*A.
3. Both S*A and S* contain parallel statements. However, because the level of parallelism in microprograms is lower than we wish to normally describe in endoarchitectures, more types of parallelism are representable in S* than in S*A.[1]
4. It was stated in Chapter 1 that the main purpose of machine architecture is that it allows us to abstract from the physical machine itself. We may, in fact, wish to view an architecture as a set of logical information processing systems that we call *architectural systems*. We need a means of designing and analyzing such systems in isolation if necessary. Thus, in the design of S*A particular attention has been given to the problem of modularizing architectural descriptions. The principal higher-level constructs for this purpose in S*A are the system and the mechanism. These constructs are absent in S*.

---

[1] Throughout this discussion and in the chapters that follow, whenever I say that something is representable or describable in S*, or talk about an S* program, it must be understood that, strictly speaking, I am referring to an instantiated version of S*. Since S* is a schema, one cannot write a program or describe anything using S* alone.

DATA TYPES AND DATA OBJECTS IN S*A

In the sections that follow, I shall provide an informal and partial description of S*A. For a more rigorous definition of the language, the reader must refer to the S*A report, reproduced in this monograph as Appendix A.

5.3 DATA TYPES AND DATA OBJECTS IN S*A

The primitive data type in S*A is the **bit**, consisting of the values [0,1]. All other types are structured from this primitive. Given data objects of type **bit**, the usual arithmetic and Boolean operations are defined on them.

Structured data types are of five categories:

1. The sequence:

$$\textbf{seq}[a..b]\textbf{bit} \qquad (5.1)$$

where a, b (a ≥ b) are integers that define the high- and low-order bounds of the bit sequence, respectively.

2. The array:

$$\textbf{array}[a..b]\textbf{of } T \qquad (5.2)$$

3. The stack:

$$\textbf{stack}[i] \textbf{ of } T \textbf{ with } V \qquad (5.3)$$

where i is an integer specifying the maximum depth of the stack, T is the type of the stack elements, and V is the name of a variable that acts as the stack pointer. The **with** clause is optional; when specified, however, it restricts the use of a stack pointer to V.

The obvious characteristics of a **stack** type data object is that only its "top" element is accessible. The data type carries with it the standard primitive operations [push,pop].

4. The tuple:

$$\begin{array}{l}\textbf{tuple } V_1 : T_1 \\ \phantom{\textbf{tuple }} V_2 : T_2 \\ \phantom{\textbf{tuple }} \ldots \\ \phantom{\textbf{tuple }} V_n : T_n \\ \textbf{endtup}\end{array} \qquad (5.4)$$

The **tuple** in S*A corresponds to the Pascal **record** or the Algol 68 **struct**, and consists of a number of components or fields, with identifiers $V_1, \ldots, V_n$ of types $T_1, \ldots, T_n$ respectively. Valid operations are only defined on the individual fields (providing that they, in turn, are not tuples), and are determined by the field type.

5. The associative array:

$$\textbf{assoc array } [a..b] \textbf{ of } T \qquad (5.5)$$

where a, b (a < b) denotes the bounds of the array and T specifies the type of the array elements. Unlike the **array** type, an **assoc array** is accessed associatively. For example, given a store declared as:

assoc array [0..31] **of tuple**
                address : **seq** [15..0] **bit**;
                value : **seq** [15..0] **bit**
        endtup

the value field of particular words may be accessed by an associative match between the address fields and some other data object.

Example 5.1

1. **tuple** opcode : **seq** [6..0] **bit**;
      operand : **seq** [14..0] **bit**;
      indirect : **bit**
endtup
2. **array** [0..15] **of** register;
3. **stack** [8] **of seq** [15..0] **bit with** stack_ptr.

In using S*A for the description of architectures, the designer may, of course, freely define data objects as required, just as a programmer introduces variables into a program. More precisely, the declaration of a data object of a given type T in an S*A description denotes the existence of a storage device whose abstract properties are prescribed by the properties of the data type T, in the architecture being designed or, if it already exists, being described.

A data object is always defined as being either *global* (**glovar**) or *private* (**privar**) relative to the mechanism (i.e., the basic unit of architectural description) or the system in which it is defined. This aspect is further elaborated when I discuss the mechanism construct.

Example 5.2

type register, word = **seq** [15..0] **bit**;
glovar main_mem : **array** [0..4095] **of** word;
privar local_store : **array** [0..31] **of** register

DATA TYPES AND DATA OBJECTS IN S*A 45

In addition to variables, the architect may also need to declare and use predefined constants. These represent values of read-only storage devices.

Example 5.3

    const   zeros : **dec** (16)0,
               plus1   : **dec** (16)1,
               sign   : **bin** (8)10000000;

Here, the integer in parentheses gives the length of the data object in bits, and the rightmost number its (invariant) value in the number system indicated by the keyword **dec, bin, oct,** or **hex.** Thus the first line in the above example declares a constant data object named "zeros" which has a length of 16 bits and a (decimal) value of 0.

In addition, S*A allows pseudoconstant declarations which, apart from beginning with the keyword **pconst,** have the same syntax as constant declarations. Pseudoconstants simply allow names to be associated with declared constants for the purpose of textual clarity. For example, given the declaration:

$$\textbf{pconst } \text{indir} = \textbf{dec}(4)8$$

the name "indir" may be used instead of the literal 8 in an executional statement. Unlike constants, pseudoconstants do not denote a hardware-defined data object in the implementation of the architecture.

S*A also incorporates the concept of a *channel,* which is a special class of data objects that provide a means of representing data communications between mechanisms. Informally, the S*A channel may be regarded as an abstraction of communication paths (e.g., buses), like the line variable in the I/O description language SLIDE. The channel declaration is of the form:

$$\textbf{chan } \text{identifier}: \textit{type}$$

where "type" is a **bit** or a sequence. In S*A, the main distinction between channels and variables are:

1. Variables are abstractions of storage objects, while channels represent communication paths.
2. Variables may be private or global to a mechanism, while channels are always global.
3. State transitions in channel bits are *testable conditions*—that is, transitions from 0 to 1 or 1 to 0 may be tested by using special expressions like ($0 < x < 1$ or $1 > x > 0$, where x is an identifier for a particular channel bit). Finally, a channel can only be of type **bit** or **seq.**

5.4 THE SPECIFICATION OF ACTION IN ARCHITECTURAL DESCRIPTIONS

S*A includes most of the major constructs present in Pascal-like languages for the specification of actions. The basic construct is the *assignment statement,* usually of the form:

$$x := E \tag{5.6}$$

where x identifies a data object and E is an expression. In S*A, evaluation of an expression denotes the activation of a combinational circuit that accepts as inputs the values in the specified objects named in the expression, and produces an output value according to the operators specified in the expression. This value is assigned to the data object x.[2]

Remark. The assignment statement provides an example of how the architectural description level may differ from the register-transfer level. In the latter, additional data and assignment types that are of no consequence at the more abstract architectural level may be necessary for the specification of information transfer—for example, data types that represent classes of terminals (rather than variables), and an assignment type that represents interconnection (of terminals) rather than information flow (between variables).

The assignment statement is supported by a set of control statements (see Table 5.1). These include conventional constructs for repetition, namely:

$$\textbf{repeat } S \textbf{ until } B \tag{5.7}$$

$$\textbf{while } B \textbf{ do } S \textbf{ od} \tag{5.8}$$

and the two selection statements

$$\begin{aligned}&\textbf{if } B_1 => S_1 \\ &\| \ B_2 => S_2 \\ &\ \ldots \\ &\| \ B_n => S_n \\ &\textbf{fi}\end{aligned} \tag{5.9}$$

and

$$\begin{aligned}&\textbf{case } V \textit{ of} \\ &L_1 : S_1 \\ &L_2 : S_2 \\ &\ \ldots \\ &L_n : S_n \\ &\textbf{endcase}\end{aligned} \tag{5.10}$$

---

[2] In its most general form there may be a list of identifiers on the left hand side of the statement;

In (5.7) and (5.8), B denotes a Boolean expression and S an executable statement. In (5.9) each B_i designates a Boolean expression defined over a subset of declared data objects. If B_i is true then the corresponding statement S_i is selected for execution.[3] The **if** statement is strictly deterministic in that at most one of the B_i's can be true at any time.

In the **case** statement (5.10), V is a (**seq** type) data object, and $L_1, \ldots, L_n$ are integers such that V may take on these, and only these, values. In executing this statement V is evaluated; if the state of V is L_i the corresponding statement S_i is selected for execution.

A special case of the **while** construct (5.8) must be noted. This is the nonterminating repetition:

$$\textbf{forever do } S \textbf{ od} \quad (5.11)$$

which is formally equivalent to the statement:

$$\textbf{while } \text{true} \textbf{ do } S \textbf{ od}$$

Several control statements are associated with the activation and deactivation of S*A procedures (see Sections 5.5 and 5.6). The statement

$$\textbf{call } p \quad (5.12)$$

initiates the named procedure p and transfers control to it. The **call** terminates only when control is returned from p, and the statement following the **call** begins execution only after the **call** has terminated.

The activate statement:

$$\textbf{act } p \quad (5.13)$$

initiates the named procedure p and terminates. The statement following the activate statement then begins execution, in parallel with p.

The statement

$$\textbf{trap } p \quad (5.14)$$

---

that is, $x_1, x_2, \ldots, x_n := E$. In this case the value of E is broadcast to all the data objects $x_1, x_2, \ldots, x_n$.

[3] From a pragmatic point of view, it must be noted that an assignment (or any other executable) statement in a machine description language is a direct specification of action in the machine hardware. This is in contrast to an executable statement in a programming language, which is implicitly effected by an underlying (software, firmware, or hardware) interpreter. For convenience of terminology, however, we shall continue to use the terms *execution* and *executable statement* in the description of such S*A statements that specify action.

is basically identical in action to the procedure **call** except that control may not return from p to the procedure that originated the trap. In other words, a **trap** statement may never terminate.

The statement

$$\textbf{exit} \qquad (5.15)$$

causes the procedure containing the statement to terminate. That is, control is explicitly forced to the end of the procedure. Finally, the

$$\textbf{return} \qquad (5.16)$$

statement within a procedure p causes control to return to the procedure that originally invoked p. Thus **return** is a means of explicitly forcing a return of control from the invoked to the invoking procedure. Note that in the absence of a **return,** a procedure p that has been called would automatically return control to the parent procedure when the control reaches the end of p.

Examples of the uses of these constructs will be given in later sections and chapters.

5.5 MODULAR DESCRIPTIONS

One of the earliest organizational concepts the student of programming and computer design is introduced to is that of *modularization*. In the design of hardware systems this concept is perhaps most closely associated with the physical composition of the system. In particular, with developments in large scale integration (LSI) and very large scale integration (VLSI) technology, rather complex physical modules may serve as the primitive building blocks for constructing computer systems—for example, the uses of register-transfer level modules (RTMs) or a family of bit-slice logic devices in developing conventional uniprocessor systems, or the synthesis of computer networks in which an entire microprocessor serves as the basic module.

In the software domain, modularization serves many purposes. The best established of these is to provide a means of structuring programs, thereby reducing the complexity of both the program and the design process. In addition, modularization provides an important means for program maintenance and portability. In recent years some fresh insights have been added on this issue, most notably those of *information hiding* and *data abstraction*.

In presenting S*A, we consider first the notion of *modular architecture descriptions,* the purposes they may serve in design, description, and realization of computer architectures, and the relationship of such descriptions with software modules on the one hand and hardware modules on the other.

An architecture of any reasonable complexity can be regarded as a collection of abstract, functionally distinct information processing systems. I shall

MODULAR DESCRIPTIONS

call these *architectural systems;* we may find a need for describing, evaluating, and studying such entities in isolation.

Example 5.4

In the PDP-11, the addressing modes and the autoincrement/autodecrement feature are examples of interesting exoarchitectural systems. In the MU5, the (primary and secondary) instruction pipelines and the branch–look ahead feature are nontrivial systems at the endoarchitectural level. The residual control feature and the two-level instruction interpretation scheme in the QM-1 provide examples of microarchitectural systems.

In S*A, the *system* construct is provided for this purpose. A system in S*A describes an integrated interacting set of components that bear some particular relationship to another. The components themselves may constitute simpler systems. Ultimately, the simplest form of the construct will describe some set of primitive architectural modules that satisfy the following requirements:

1. It should be possible to describe architectural modules that act as critical regions.

   Example 5.5

   A main memory can be accessed by two operations, READ and WRITE. When the memory is shared between two or more processors, only a single READ or WRITE operation can be going on at any one time.

2. The architect should be able to specify architectural data abstractions—that is, a collection of data types together with the operations defined on them—as single entities.

   Example 5.6

   Consider a capability store. This is an associative memory, each word of which holds a capability with the following format:

SEGMENT ID	BASE ADDRESS	LENGTH	ACCESS RIGHT

   Defined on this capability store are a pair of operations that we may call FINDADDR and LOADCAP; the former searches the store for an entry corresponding to some SEGMENT ID and returns the contents of the word if such an entry exists. LOADCAP causes a new capability to be entered into the capability store, either in an empty word or by overwriting an existing entry. Both these are hardware operations.

The capability store and its operations form an architectural data abstraction. The utility of such abstractions lies in essentially two directions. First, in the context of a design methodology, by encapsulating the store and its operations within a single descriptive module, an appropriate physical module is suggested for the implementation phase. The resulting capability unit could consist of the store, matching circuitry, internal registers, and data paths corresponding to the data abstraction. The control logic could also be a part of the unit rather than a central control unit. Thus, there would be a natural match between the architectural and physical modularization of the system.

Second, by enforcing the rule that the only operations defined on the capability store are those specified within the data abstractions, we provide for clean, well-defined specifications and restricted (controlled) access to the store. This enhances the verifiability of the design.

3. The modularization construct should facilitate information hiding. Taking the capability example once more, suppose we describe it such that from the "outside" the capability store is not even visible. It should then be possible to specify the behavior of the module purely in terms of the state changes in its interface data objects, without any reference to the store itself. The great advantage of this is that it facilitates changes in the design of the hidden object, with minimal impact on other aspects of the architecture. Indeed, the architect may decide to postpone decisions concerning the design of the capability store.

4. In addition to the above requirements, the construct should allow more conventional decompositions to be effected. That is, for enhancing clarity and ease of understanding, it should provide a simple yet adequate means of segmenting an architectural system into smaller, simpler modules. Conversely, a particular component of the architecture may not naturally be able to be composed into a system, and yet be of sufficient complexity to merit a module of description of its own.

5.6 MECHANISMS AND SYSTEMS

The **mechanism** construct in S*A provides a means for describing primitive architectural modules such that the four requirements discussed above are met. In its most usual form, an S*A **mechanism** (Fig. 5.1) consists of global

```
Global state variables (glovars)
Private state variables (privars)
Public proceduces
Private procedures
```

Figure 5.1 Components of an S*A mechanism.

state variables (glovars), private state variables (privars), public procedures (which can be invoked from other mechanisms), and private procedures.

A mechanism is quasisequential in that while each of its constituent procedures is a statement sequence, the statements may themselves be composed of more elementary statements executing concurrently in a simple, synchronous (clocked) fashion. This form of concurrency is denoted by the statement form:

$$S_1 \square S_2 \square \ldots \square S_n \qquad (5.17)$$

which designates that $S_1, S_2, \ldots, S_n$ begin execution together. Any statement that follows begins execution only when $S_1, S_2, \ldots, S_n$ have all terminated.

A procedure in S*A is a specification of the data flow that is necessary for realizing a particular hardware function. The data flow implies an underlying data path in the physical computer, the structure of which, however, is not explicitly definable in S*A.

A mechanism is *active* whenever one of its procedures is active. A procedure is active from the time it is called to the time it returns control. Mechanisms may be activated by *initiation statements* (which correspond to the situation where the machine is started up) or by calls on their public procedures from public procedures inside other mechanisms. Public procedures inside the same mechanism cannot call one another.

Private procedures can only be invoked from other procedures inside their own mechanism. Thus, the only ways one mechanism may interact with another are the invocation of one mechanism's public procedures by another's, by exchanging synchronizing signals (see Section 5.7), or by accessing common (shared) global variables.

Mechanisms satisfy a mutual exclusion rule stating that a public procedure cannot be activated when its surrounding mechanism is already active.

Mechanisms are organized into larger logical entities called *systems*.

The reader may take note of how the semantics of the mechanism construct meets the requirements prescribed in the previous section. The rule that public procedures can be invoked only in a mutually exclusive fashion ensures that *critical regions* can be specified. By encapsulating data storage objects together with associated operations (procedures) in a single mechanism, the architect may define *data abstractions*. Finally, by distinguishing global variables and public procedures—entities that are visible to the mechanism's environment—from private variables and procedures—entities that

```
/* mechanism for accessing data stack memory */
mech DSTK_ACCESS;
   type dstk_mem_word = tuple
                          .....
                        endtup;
   privar dstk_mem : array[..] of dstk_mem_word;
                        /* stack memory */
   glovar dstk_ptr : seq[..] bit
                        /* stack pointer */
   glovar dstk_buffer : dstk_mem_word;
                        /* source or destination for stk opn */
   glovar dstk_mem_ar : seq[..] bit;
                        /* random access register */

   proc PUSH;
      /* push contents of dstk_buffer onto stack */
      if dstk_ptr = MAX => trap STACK_OVFLOW fi;
                        /* MAX is some constant */
      dstk_mem [dstk_ptr] := dstk_buffer;
      dstk_ptr := dstk_ptr + 1
   endproc

   proc POP;
      /* pop top of stack into dstk_buffer */
      dstk_ptr := dstk_ptr - 1;
      if dstk_ptr = MIN => trap STACK_UNDFLOW fi;
                        /* MIN is some constant */
      dstk_buffer := dstk_mem [dstk_ptr]
   endproc

   funct VAL_AT_DSTK_MEM_AR;
      value : dstk_mem_word;
      value := dstk_mem [dstk_mem_ar];
   endfunct
endmech
```

Figure 5.2 S*A description of a stack memory (copyright © 1982, Academic Press; reprinted with permission).

are not—the concept of *information hiding* is made available to the architect/designer.

Example 5.7

Consider an S*A description of a small component of the student PL-Machine (SPLM) exoarchitecture, namely, the manipulation of a data stack (Myers [1982]). Since stack manipulation may be required as a component in the execution of several instructions, we would like to represent the stack along with a general set of operations that may be invoked during the execution of relevant instructions. Thus we might describe the stack along with its operations as a single mechanism, as shown in Fig. 5.2.

Remarks

1. The variable dstk_mem is declared here as private to the mechanism. Hence the only means by which any other mechanism may access it is

by calling one of its public procedures (including the function procedure). The latter (viz., PUSH, POP, VAL_AT_DSTK_MEM_AR), along with the three global variables, provide the mechanism's interface with its environment.

2. The **trap** statements, when executed, result in the invocation of private procedures STACK_UNDFLOW and STACK_OVFLOW (not shown here).

3. In some cases, the range bounds of a sequence or an array are given as [..]. This is syntactically legal in S*A and indicates that the bounds exist but need not be specified at this stage of design.

Example 5.8

A mechanism may be shared between systems. For example, in SPLM, there are several instructions whose executions require manipulation of dstk_mem. Two of these are SNAME and EVAL; both are sufficiently complex to merit descriptions as individual systems. Figure 5.3 shows, for instance, the system description for SNAME; it consists of two mechanisms, SNAME_OPN and DSTK_ACCESS. The instruction EVAL can be similarly structured into two mechanisms, EVAL_OPN and DSTK_ACCESS, the former invoking the latter. Figure 5.4 shows a skeletal outline for this system. Since a mechanism can only be activated serially, DSTK_ACCESS behaves as a *monitor* with respect to the two systems. Note that Fig. 5.3 also provides an instance of a synchronous parallel statement.

Whenever two or more mechanisms within a system share a set of common data objects, these may, for convenience, be declared at the beginning of the system declaration rather than repeated inside the mechanisms. The scope of such a declaration is, therefore, the entire system containing it— that is, all the mechanisms encapsulated by the system.

Note that whether a data object is global or private is determined strictly by the mechanism or system which immediately contains it. Thus, for example, if data objects $D_1, \ldots, D_4$ are common to mechanisms M_1, M_2, and M_3, and these mechanisms are encapsulated in a single system declaration, then the data objects may be described in two ways. If they are declared at the beginning of a system S, they will appear as private data objects (with respect to S) as shown by (5.18):

$$
\begin{aligned}
&\textbf{sys } S; \\
&\quad \textbf{privar } D_1, D_2 : \textbf{seq } [..] \textbf{ bit}; \\
&\quad \textbf{privar } D_3, D_4 : \textbf{tuple} \ldots \textbf{endtup}; \\
&\quad \textbf{mech } M_1 ; \ldots \textbf{endmech} \\
&\quad \textbf{mech } M_2 ; \ldots \textbf{endmech} \\
&\quad \textbf{mech } M_3 ; \ldots \textbf{endmech} \\
&\textbf{endsys}
\end{aligned}
\qquad (5.18)
$$

```
sys SNAME;
    mech SNAME_OPN;
        glovar inst_reg : tuple high : seq[7..0] bit;
                                low  : seq[7..0] bit
                          endtup;
        glovar display_reg : array[0..15] of seq[7..0]bit;
        glovar dstk_ptr : seq [..] bit;
        glovar dstk_buffer : seq [38..0] bit;
        pconst indir = dec (4)8;


        proc UPDT_STACK;
            do dstk_buffer[31..20] :=
                        display_reg[inst_reg.low[7..4]]
             ¤ dstk_buffer[19..0] := inst_reg.low [3..0]
             ¤ dstk_buffer[38..36] := 0
             ¤ dstk_buffer[35..32] := indir
            od;
            call DSTK_ACCESS.PUSH;
                    /* push "dstk_buffer" onto "dstk_mem"*/
        endproc
    endmech;


    mech DSTK_ACCESS;
        . . . . .
    endmech
endsys
```

Figure 5.3 Outline of an S*A system (copyright © 1982, Academic Press; reprinted with permission).

In this case, $D_1, \ldots, D_4$ will not appear as declarations inside M_1, M_2, and M_3. However, it is implied that $D_1, \ldots, D_4$ are global with respect to the mechanisms.

Alternatively, one may define the system S according to (5.19), in which case each of the mechanism declarations would contain explicit declarations of $D_1, \ldots, D_4$ as global data objects; that is, $D_1, \ldots, D_4$ are global with respect to the mechanism containing them.

```
sys EVAL;
         mech EVAL_OPN; . . . . . endmech
         mech DSTK_ACCESS; . . . . . endmech
endsys
```

Figure 5.4 A skeletal S*A system (copyright © 1982, Academic Press; reprinted with permission).

```
sys S;                                              (5.19)
    mech M₁;
        glovar D₁, D₂ : seq [..] bit;
        glovar D₃, D₄ : tuple . . . endtup
        . . . . . . .
    endmech;
    mech M₂ ; . . . endmech;
    mech M₃ ; . . . endmech
endsys
```

Figures 5.2 through 5.4 describe part of an exoarchitecture. In general we can specify the entire exoarchitecture of SPLM (or of any other machine) as a single S*A system, its precise nature being determined by the amount of detail we wish to include. For instance, the overall SPLM machine can be organized in the form of Fig. 5.5. Note that a system may contain other embedded systems.

5.7 ASYNCHRONOUS CONCURRENT SYSTEMS

High-performance computer systems usually have some sort of parallelism or concurrency at the endoarchitectural level, and hence an architectural

```
sys SPLM;
    mech INST_FETCH; . . . . . endmech;
    sys INST_EXEC;
        sys SNAME;      . . . . . endsys;
        sys LNAME;      . . . . . endsys;
            . . . . .
        sys EVAL;       . . . . . endsys
    endsys
endsys
```

Figure 5.5 An S*A system with other embedded system.

language must have some capability for the representation of parallel actions. Note that I have already introduced one form of concurrency—the □ operator, which allows the architect to specify the execution of a set of statements in a synchronous manner (5.17).

We need, in addition, constructs for describing systems of concurrent asynchronous mechanisms. Such systems will be called *asynchronous systems*.

The basic concept in concurrent system design is that of synchronization. S*A provides a semaphore-like abstraction called the *synchronizer* consisting of **bit** and **seq** type state variables on which the operations **await** and **sig** are defined. Thus, given the declaration

$$\text{sync } X : \text{bit}$$

or

$$\text{sync } X : \text{seq } [..] \text{ bit}$$

the statement "**await** X" will never terminate as long as the integer-valued state of X is 0. Whenever X > 0, the **await** operation decrements the value of X by one and terminates. The statement "**sig** X" is only defined if the integer-valued state of X is less than MAX, the maximum integer-valued state that X may assume. In that case, the effect of the **sig** operation is simply to increment X by one.[4]

A synchronizer is a primitive critical region in that only one **await** or **sig** operation can be operating on it at a time. Note also that synchronizers must, by their very nature and purpose, be global variables relative to a particular pair of mechanisms.

The **await** operation can also be used in the form

$$\text{await } X \text{ do } S \text{ od}$$

Here, the statement S will begin execution if and only if the substatement "**await** X" terminates. This is functionally equivalent to the statement

$$\text{await } X; S$$

except that the former is *indivisible*—that is, one cannot jump to a statement inside S from outside.

---

[4] We may define the semantics of the **await** and **sig** operations more formally using Hoare-type assertions of the form {P} S {Q}, which is to be read as: if the assertion P holds prior to the execution of S, then when S terminates the assertion Q will hold. Accordingly we obtain the following semantic rules for the synchronizing operations:

$$\text{await } X : \{X = X_0 \land X > 0\} \text{ await } X \{X = X_0 - 1 \land X_0 \geq 0\}$$
$$\{X = X_0 \land X = 0\} \text{ await } X \{\text{false}\}$$
$$\text{sig } X : \{X = X_0 \land 0 \leq X < \text{MAX}\} \text{ sig } X \{X = X_0 + 1 \land X \leq \text{MAX}\}$$

Here, "false" is the predicate that is always false.

Concurrent systems may consist of mechanisms that, once turned on, are forever active. Once such a mechanism is activated it will remain so, providing its (nonterminating) procedure does not contain a goto command to some point outside the procedure. In its physical realization, a nonterminating procedure may be initially activated whenever the machine is started up. In S*A this is described by *initiation* statements of the form:

init X.a

where X and a are names of the mechanism and a public procedure within it, respectively.

I shall defer giving an example of asynchronous systems until Chapter 7, where an asynchronous instruction pipeline is described in some detail.

5.8 SYNONYMS

In S*A, variables represent hardware-implemented storage objects. In designing architectures a need often arises for (1) associating with the same storage object either alternate data structures or names or both; and (2) describing software-generated data structures residing within main memory.

Example 5.9

In the SWARD machine exoarchitecture (Myers [1982]), the principal storage object, termed a *module,* is a data structure consisting of three components, the *module header, address space,* and *instruction space*. The header defines certain attributes of the module, while the address space consists of a series of *cells,* containing the data accessible to the module. The instruction space contains a sequence of machine instructions.

Thus, in the SWARD machine, main memory may be viewed both as a linear array of words and as a collection of modules, varying in size but all possessing the basic structure just outlined. In describing the semantics of SWARD instructions, the system's addressing characteristics, and other exoarchitectural features, the structure of these modules must be taken into account.

In S*A, given a global or private variable declaration, the architect may provide such alternative types and names through the *synonym* declaration:

syn "synonym identifier" : "type" = "object identifier"

where the synonym identifier is the new name, the type (an optional clause) is the associated data type, and the object identifier identifies a previously declared data object (or part of it) or another synonym. Clearly, the type associated with the synonym must be consistent with the type of the object referenced on the right-hand side of the declaration.[5]

Example 5.10

Given a variable X declared as

$$\textbf{glovar } X : \textbf{array } [0..63]$$

a synonym may be defined through a declaration. For example,

$$\textbf{syn } y = X[0..7]$$

This is a static declaration, in the sense that the name "y" will always refer to the subarray X[0..7]. This may be contrasted to the declaration

$$\textbf{syn } Z = X[V_1..V_2]$$

where V_1 and V_2 are both variables. In this case, the subarray corresponding to the synonym Z varies and will be determined dynamically—whenever Z is referenced in the procedural part of some mechanism, the actual subarray of X being accessed is determined by the values in V_1 and V_2 at that particular time. Note that the optional type clause has been omitted in both these declarations.

Example 5.11

In the case of the SWARD example, the data type "module" may now be declared along the following lines (Fig. 5.6). On the one hand, the synonym statement defines "curr_mod" as an instance of the data type "module"; on the other, it maps this object into the array "mainmem." Hereafter one may refer to "curr_mod" as if it were an independent (tuple type) data object. Whenever such a reference is made, precisely which part of "mainmem" is being addressed is determined by the values of the variables "curr_mod_addr_" and "curr_mod_end_addr" at the time.

5.9 MECHANISM TYPES

Recall that a mechanism represents a set of state variables together with one or more procedures that operate on them. Although the mechanisms pre-

---

[5]In the most general form of the declaration, *several* alternative types may be specified for a synonym in the same declaration, just as *several* alternative types may be specified in a variable declaration. Each type designates a distinct structure associated with the same name. Examples of such declarations will be given in a later chapter.

MECHANISM TYPES

```
type token = seq[3..0] bit;
                              /* basic storage unit in SWARD */
type module =
   tuple header : tuple ... endtup;
         address_space : tuple ... endtup;
         instr_space : array [ ] of ...
   endtup;
glovar mainmem : array [..] of token;
glovar curr_mod_addr, curr_mod_end_addr : seq [..] bit


syn curr_mod : module =
         mainmem[curr_mod_addr..curr_mod_end_addr]
```

Figure 5.6 Data type clause in an S*A synonym statement (copyright © 1982, Academic Press; reprinted with permission).

sented so far have all described unique objects, it is conceivable that a machine description may need two or more mechanisms that are identical to one another in structure. That is, a given machine description may need to include several instances of the same mechanism. This leads to the notion of defining mechanism types and declaring mechanisms as instances of such types.

Consider as an example the system depicted in Fig. 5.7 (Leung [1979]). Here, a number of arithmetic-logic units (ALUs) communicate with a controller—the latter consisting of two mechanisms (which we may call IN and OUT) that are responsible, respectively, for receiving "operation packets" from, and sending "result packets" to, the outside world. In addition the system contains a collection of independent "drivers," one for each ALU.

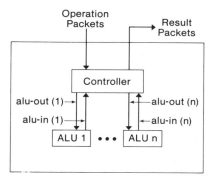

Figure 5.7 Multiple ALU-controller complex.

```
sys DRIVERSYSTEM;
   type mech ALU_DRIVER =
      sync alu_free : bit(1); /* 1 implies alu is free */
      privar opt : opn_pkt; /* holds operation packet */
      . . . . .
      proc DRIVE_ALU;
         forever do
            await alu_free
               do opt := IN.GET_OPN_PKT;
                  /* send opt to ALU &
                        receive result from ALU */
                  . . . . .
               od;
            sig alu_free
            /* driver sends out "tagged" result to
                              various destinations*/
               . . . . .
         od
      endproc
   end type mech
   sys DRIVER : set(1..n) of ALU_DRIVER;
   init DRIVER(1).DRIVE_ALU,..,DRIVER(n).DRIVE_ALU
endsys
```

Figure 5.8 ALU driver system in S*A.

Structurally and operationally, the ALU drivers are identical to one another. We should therefore be able to describe the drivers concisely by declaring a mechanism type called "ALU_DRIVER" and defining a set of instances of such a mechanism. Figure 5.8 shows a part of this description. The mechanism type ALU_DRIVER is defined exactly as a mechanism would be. Subsequently, a system called DRIVER is declared as a set of n instances of ALU_DRIVER. The result of this declaration is that the mechanisms in DRIVER are automatically identified uniquely as DRIVER(1), DRIVER(2), . . . , DRIVER(n). Further, each data object identifier X appearing in the mechanism type definition is automatically known in

DRIVER(*i*) as *X*(*i*) (for $1 \le i \le n$)—for example, alu_free(1), result(1), opt(1) lie within DRIVER(1).

Note that DRIVERSYSTEM will be part of a larger concurrent system representing the controller that will also include mechanisms for IN and OUT and mechanisms representing the *n* ALUs. The procedure DRIVE_ALU is nonterminating, as indicated by the **forever** statement. All *n* copies of this procedure (one in each DRIVER(*i*), i = 1, . . . , *n*) are initially activated by the **init** statement. Note also that in S*A, variables can be initialized to describe values by specifying the value as part of the declarations—for example, "alu_free" is initialized to 1.

5.10 BIBLIOGRAPHIC NOTES

The microprogramming language schema S* has been described in several papers by Dasgupta (1978, 1980a, 1980b), while Klassen and Dasgupta (1981) provides details of an instantiation of S* with respect to the Nanodata QM-1. The language S*A was first presented in Dasgupta (1981). For a preliminary discussion and rationale of the S* family see Dasgupta and Olafsson (1982).

The S*A language represents one particular approach to the design of architecture description languages. Several other language proposals have emerged in the last decade, most notably ISPS in Barbacci (1979, 1981) and Barbacci and Northcutt (1980), ADL in Leung (1979, 1981), and SLIDE in Parker and Wallace (1979, 1981). The CONLAN project is an ambitious effort to construct an extensible family of hardware description languages, including the description of architectures; it is described in Piloty et al. (1980a, 1980b, 1980c). The full language definition is given in Piloty et al. (1981). Dasgupta (1982) discusses and compares these and other languages.

The Student PL machine (SPLM) mentioned in this chapter was originally due to Wortman (1972). The SWARD machine was designed by Myers and is discussed in Myers (1982). SPLM is also described by Myers. The multiple ALU/controller complex discussed in Section 5.9 is based on the ALU controller described by Leung (1979).

SIX

FORMAL DESIGN OF A MICROCODE LOADER

6.1 THE PROBLEM

The quintessential architectural problem is one where the architect is required to produce a logical or abstract form that will perform a required information processing task. The form must also satisfy some given set of constraints—including the fact that it is to be implemented in hardware and/or firmware.

The form in question may, for example, be an addressing structure, a descriptor architecture, or an instruction class, at the level of exoarchitecture; or a cache system, an instruction pipeline, or an address translation mechanism at the endoarchitectural level. In the most general case, the entire computer architecture is the form in question. But whether we consider the whole or its parts, the goal remains common: the conditions, constraints, and requirements that collectively give birth to the architectural problem originate in the user or problem solving environment. The computer architect mediates between this environment and the universe of hardware devices.

Having established one of the basic tools required for formal architectural design—a description language—we would now like to apply it to a problem. At this stage it is not my objective to construct a design methodology, or to design a system from first principles. Rather I shall consider an already solved problem—an informally designed subsystem—and construct its description in S*A. I shall then use the semantic rules of S*A to verify that the description does indeed satisfy the original requirements and thus provide a partial demonstration that we are on the right track toward a formal design process.

Consider a microprogrammed machine with several sets of instructions. One is a general set while the others are special sets tailored to the requirements of particular programs or languages. The microroutines for the general instruction set reside permanently in (writable) control store, but those for the special sets are loaded from main store only when required. This discussion is concerned with the loading problem.

The problem was first posed and solved by Wilkes (1973) in the context of providing hardware support for multiple-language-directed architectures. By "solved" I mean that Wilkes provided an architectural form that behaves in the required fashion and satisfies, at the same time, certain constraints. The solution was architectural in that the form was designed for implementation in hardware and firmware.

As stated by Wilkes, the loading mechanism had to satisfy the following requirements:

[R6.1] Programs that use the general instruction set only should require no more time to execute than they would in the absence of the loading mechanism.

[R6.2] If the microcode for a special instruction is already loaded, that instruction should run as fast as a regular instruction.

[R6.3] The algorithm for reusing control memory should be effective and fast.

[R6.4] There should be as few restrictions as possible on sharing microcode between instructions.

Remarks

1. Whether a given loading mechanism complies with the requirements R6.1, R6.2, and R6.4 can be determined from the design itself. These may, then, be said to be *design time* requirements. In contrast, R6.3 is at least partially an *implementation time* requirement. However, it has become standard practice for algorithm designers to analyze and determine the asymptotic efficiency of their algorithms. Such time and space measures are typically expressed as $O(f(n))$ where $f(n)$ is a function of n, an appropriate parameter of the input to the algorithm. It is thus possible to assess the efficiency of an algorithm without having to implement it, or to evaluate the efficiency of a program without executing it. In this sense, R6.3 is, in part, a design time requirement.

2. Note that we may not be able to anticipate whether or not the requirements are consistent with one another. Following Alexander (1964)

there may, indeed, be a misfit between requirements, in the sense that satisfying all or even a subset of the requirements may simply not be possible. In other words, we may not be able to arrive at an optimal design. Simon (1981) has suggested that in solving problems of design or in making decisions, the designer must frequently be content to "satisfice" rather than optimize—that is, seek a good solution rather than the best one.

In anticipation of the possibility of mismatching requirements, and the consequent need to satisfice the design, the following (meta)requirement may be added:

[R6.5] R6.1 and R6.2 are the strongest requirements in the sense that the design must satisfy these. R6.3 then follows. R6.4 is the weakest requirement.

6.2 AN INFORMAL DESIGN

Let us first consider the informal design as proposed by Wilkes (1973). A formal version of this design can then be developed in S*A. Note that the prior existence of the informal design does not detract from the value of having a formal version. The latter serves as a more rigorous characterization of the architecture than the original proposal; it allows us to demonstrate that the form is correct vis-à-vis the requirements—in particular R6.1 and R6.2; and finally, the architecture can be more clearly conceived in this form than in its less formal incarnation.

Each set of special instructions is identified by a unique instruction set number (ISN) and is defined by a microcode table residing in main memory. The table contains the microroutines in binary form, ready for loading into control store. Each microroutine corresponds to an opcode. The microroutines in the table are preceded by an index specifying the location relative to the beginning of the table and the length of the corresponding microroutine. Thus, given the starting address of a microcode table, a microloader (permanently resident in control store) has all the information necessary to load the microroutine into control store. The tables' starting addresses are given in an ISN table which gives for each ISN a unique identifier of the corresponding microcode table and its starting address in main memory. At any point of time, when a program is in execution, an ISN register shows the current instruction set number.

In the Wilkes (1973) scheme, two control stores, CSI and CSII, are used (Fig. 6.1). The store CSII has a sufficiently wide word to accommodate, in addition to microinstructions, the ISN and a *validity bit*. This store is used to hold the first microinstructions of the routines that interpret the special instructions. The store CSI is a conventional control store and contains the microloader and the microroutines for the general instruction set. In addi-

AN INFORMAL DESIGN

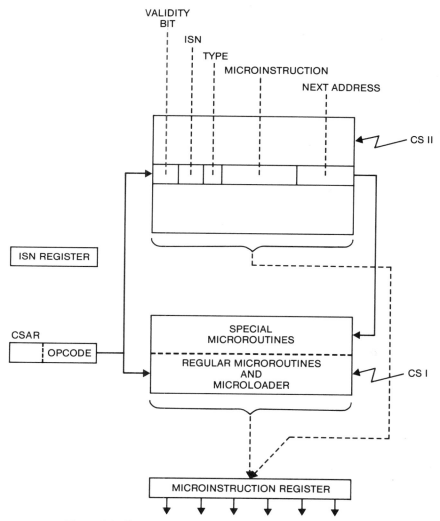

Figure 6.1 Storage components in Wilkes's loading mechanism.

tion, a part of it is used on a wrap-around basis to hold the rest of the microroutines corresponding to the special instructions.

At the start of a machine instruction's execution, an opcode is loaded into the least significant bits of the control store address register. Both CSI and CSII are then accessed. The system is to be so designed that if the opcode refers to the general instruction set then the microprogram runs wholly from CSI in the conventional way. If it refers to a special instruction set, the output of CSI is ignored and that of CSII is accepted. If the ISN field of the microword from CSII matches that of the ISN register, and the validity bit is on, the microinstruction is executed and this initiates the interpretation of the opcode. The sequencing information in the microinstruction is interpreted as referring to CSI so that the remaining microinstructions are taken from CSI.

If the match fails it means that the microroutine for the opcode has not been loaded. Execution of the current microinstruction is suspended, the first microinstruction of the relevant microroutine is loaded into CSII, the ISN bit is filled in, and the validity bits are set. The microloader in CSI then loads into the wrap-around part of CSI the remaining microinstructions, overwriting the least recently used microinstructions. The validity bit of the first microinstruction (in CSII) of any microroutine that is overwritten must also be unset.

6.3 A FORMAL DESCRIPTION

We begin by identifying the variables that will appear in the final description. At this stage we may already know certain architectural parameters. Others may not yet be known and we may not wish to make premature or arbitrary assumptions about them. However, we are assured—from a consideration of the problem specification—that these parameters will be *knowable* before implementation. Collectively, the parameters that can be known during the design phase are the *design time parameters,* in contrast to others that are the *implementation time parameters.*

Finally, there exists, in general, a third class of parameters, which I shall call *operational*. These are only known during the operation of the implemented system and may change in value over time.

Let us assume that the following parameters are known:

opcode length	$= a + 1$ bits
number of instruction sets	$= 2^{i+1}$ $(i \geq 0)$
regular instruction set number	$= 0$
special instruction set numbers	$= 1..2^{i+1} - 1$
size of control store 1 (CSI)	$= 2^{b+1}$ words $(b > 0)$
size of control store 2 (CSII)	$= 2^{a+1}$ words

The required data objects were identified at the informal design stage. Figure 6.2 describes these in the S*A notation. Note the use of the symbol [..] to describe certain array and sequence range bounds. This is a legal symbol in S*A and simply indicates that the range bounds are design time parameters that are either not yet known or decided upon, or ones that need not be specified in the formal design at all. Although the resulting architecture may be deemed incomplete, we should have no difficulty in understanding the architecture or in ascertaining its validity.

Declared variables denote hardware storage objects. But, as the informal architecture indicates, microcode tables are data structures, generated by a system routine and residing in main memory (mm in Fig. 6.2). Since the

A FORMAL DESCRIPTION

```
/* type definitions */
   type sh_mword = tuple typebit : bit;      /* short microword format */
                         microinst : seq[..] bit;
                         nextaddr : seq[b..0] bit
                   endtup
   type lg_mword = tuple validity : bit;     /* long microword format */
                         isn : seq[i..0] bit;
                         typebit : bit;
                         microinst : seq[..] bit;
                         nextaddr : seq[b..0] bit
                   endtup;
/* global data objects */
   glovar csar : tuple hiorder : seq[b..a+1] bit /* CS1 addr.reg is csar */
                       opcode  : seq[a..0] bit /* CS2 addr.reg is csar.opcode */
                 endtup;
   glovar isn.reg : seq[i..0] bit; /* current inst.set.no register */
   glovar isn_table : array[1..2^(i+1)] of /* table of inst.set.numbers */
                      tuple isn : seq[i..0] bit;
                            unique_id : seq[..] bit;
                            st_addr : seq[..] bit; /* starting addr. of
                                                      microcode*
                                             * table for this isn */
                            length : seq[..] bit /* length of microcode table */
                      endtup;
   glovar mm : array[..] of seq[..] bit;  /* main memory */
   const one : bin(1) 1;
   glovar mir : lg_mword;  /* microinstruction register */
/* private data objects */
   privar CS1 : array[0..2^(b+1)-1] of sh_mword with csar,csar.opcode;
   privar CS2 : array[0..2^(a+1)-1] of lg_mword with csar.opcode
```

Figure 6.2 Declaration of data types and data objects.

hardware loading mechanism has to have access to this data structure it becomes necessary to include a description of the microcode tables as part of the formal architecture.

Note that from an architectural viewpoint, although microcode tables reside in main memory it is not very useful to regard that particular part of main memory as simply an array of bit sequences. Rather, it is more important for the internal structure of the microcode table to be made visible. Accordingly, microcode tables will be represented by using synonym declarations.

A data type "mcode_table" is defined in Fig. 6.3 and three synonyms are declared as shown in Fig. 6.4. In particular, "curr_mcode_table" is declared

```
type mroutine = array[.] of sh_mword;
type mcode_table = tuple unique_id : seq[..] bit
                         index_part : array[] of
                             tuple opcode : seq[a..0]bit;
                                   st_addr : seq[..]bit;
                                   length : seq[..]bit;
                             endtup
                         routine_part : array[..]of mroutine;
                         table_end : bin (..) 11..1
                  endtup;
```

Figure 6.3 Description of microcode table.

to be of type "mcode_table." In fact, this last synonym declaration creates a virtual data object of one type in terms of a real data object of another type.

In Figs. 6.2 and 6.3, the brackets "[ ]" signify that the range bounds are unknown at design time but will be known at the time the mechanism is activated. The range bounds are, in other words, operational parameters. At the time the mechanism is activated the microcode tables will be in main memory, one table per instruction set number. In each such table the length of "index_part" will equal the number of opcodes in the instruction set and therefore the number of microroutines. Thus, the lengths of both "index_part" and "routine_part"—indicated in Fig. 6.3 by the brackets—will be known at operation time.

Finally, the symbol "[.]" indicates that "mroutine" is an array of variable size elements but with a maximum size limit that is known at design time. Noting that "routine_part" is an array of elements of type "mroutine," Fig. 6.3 also indicates that each element "routine_part[i]" is a variable size array of elements of type "sh_mword" up to some prespecified maximum.

The relationship among "isn_table," "mm," and "curr_mcode_table" is shown in Fig. 6.5.

```
syn mcode_table_base = isn_table[isn_reg].st_addr;
syn mcode_table_limit = isn_table[isn_reg]length;
syn curr_mcode_table : mcode_table
    = mm[mcode_table_base..mcode_table_base +
        mcode_table_limit - 1]
```

Figure 6.4 Synonym declarations.

A FORMAL DESCRIPTION

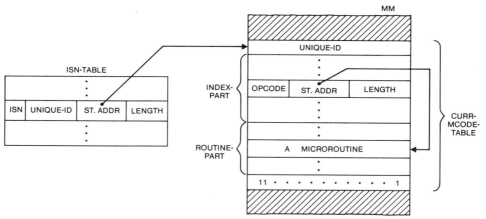

Figure 6.5 Schematic relationship between main memory, instruction set number table, and microcode table.

Remarks

1. Note that the synonym defining "curr_mcode_table" does not explicitly define all the microcode tables in "mm." It merely identifies the one that is current with respect to the currently executing instruction set.
2. In general, data objects are global or private relative to a mechanism or system. In constructing a modular description of the Wilkes (1973) scheme we are, at this stage, unsure as to whether one or more mechanisms will be required. However, we assume that since the entire design can, in any case, be encapsulated by an S*A **system** construct, the specification in Fig. 6.2 of variables as global or private is tentative, and relative to all the mechanisms within such a system.

In deciding on the modular organization of this S*A description we find, from the informal architecture, that there are two basic procedural tasks involved. Task A involves determining whether or not the microroutine corresponding to the instruction to be executed is resident in one or both the control stores and if so, initiating execution of the microinstruction. Task B applies if the microroutine is not present in either of the control stores, and involves locating the routine in main memory and loading it into the control stores in the appropriate fashion.

This may suggest a natural partitioning of the description into modules corresponding to the two tasks. From the informal design, however, it is also evident that Task A must be implemented entirely in hardware, while part of Task B (loading the routine into the control stores) will be implemented in firmware and the rest in hardware.

Thus, an alternative partitioning scheme comprising modules (possibly S*A systems) for the hardware-implemented and firmware-implemented parts may be considered. (Note that at the level of S*A description the mode of implementation is entirely transparent; it can only affect how we choose to modularize the description.)

Clearly either partitioning scheme may be followed. However, since the overall architectural description will obviously not be large, we may defer a decision on this point until it becomes clearer which is more beneficial. Instead, I shall begin with the development of a mechanism that performs Task A, expecting that the eventual shape of the overall description will emerge at a later stage. I shall call this mechanism CHECK_CTLSTORE.

6.4 A DIVERSION: FLOYD–HOARE CORRECTNESS PROOFS

Since one of the objectives of the formal design process is the verification of architectural designs—a goal that I shall pursue in the present exercise—it is useful to outline briefly the axiomatic technique pioneered by Floyd (1967) and Hoare (1969) for proving programs correct. The present section is devoted to sequential programs, while the case of parallel programs is discussed in Chapter 8.

Let P and Q be assertions about variables, and S a statement. Then the notation

$$\{P\} \; S \; \{Q\} \tag{6.1}$$

means: If P is true before execution of S then when S terminates (if at all) Q will be true. If formula (6.1) can be shown to be true we say that S is *partially correct* with respect to the assertions. If we can additionally demonstrate that S terminates, S is then said to be *totally correct*. Assertions P and Q are termed the *precondition* and *postcondition* of S, respectively.

The notation

$$\frac{P_1, P_2, P_3, \ldots, P_n}{Q} \tag{6.2}$$

means: If the assertions $P_1, P_2, \ldots, P_n$ are true then so is Q. Using such notation one may construct an axiomatic system consisting of a collection of axioms and proof rules for proving the (partial) correctness of programs. The basic axioms and proof rules for Pascal are shown in Table 6.1.

Let P be an assertion. Then the symbol P_E^X denotes P but with all free occurrences of X in P replaced by E. Thus, the axiom of assignment states that if on completion of the assignment statement X := E the assertion P holds then P (with all free occurrences of X replaced by E) holds before its execution.

A DIVERSION: FLOYD–HOARE CORRECTNESS PROOFS

Table 6.1 Basic Axioms and Proof Rules for Pascal

Rules of consequence:

(1) $\dfrac{\{P\}\ S\ \{R\},\ R \supset Q}{\{P\}\ S\ \{Q\}}$

(2) $\dfrac{P \supset R,\ \{R\}\ S\ \{Q\}}{\{P\}\ S\ \{Q\}}$

Assignment axiom: $\{P_E^X\}\ X := E\{P\}$

Rule of sequential composition:
$\dfrac{\{P_1\}\ S_1\ \{P_2\},\ \{P_2\}\ S_2\ \{P_3\},\ \ldots,\{P_n\}\ S_n\ \{P_{n+1}\}}{\{P_1\}\ \textbf{begin}\ S_1;\ S_2;\ \ldots\ ;\ S_n\ \textbf{end}\ \{P_{n+1}\}}$

Conditional rules:

(1) $\dfrac{\{P \wedge B\}\ S_1\ \{Q\},\ \{P \wedge \neg B\}\ S_2\ \{Q\}}{\{P\}\ \textbf{if}\ B\ \textbf{then}\ S_1\ \textbf{else}\ S_2\ \{Q\}}$

(2) $\dfrac{\{P \wedge B\}\ S\ \{Q\},\ P \wedge \neg B \supset Q}{\{P\}\ \textbf{if}\ B\ \textbf{then}\ S\ \{Q\}}$

Iteration rules:

(1) $\dfrac{\{P \wedge B\}\ S\ \{P\}}{\{P\}\ \textbf{while}\ B\ \textbf{do}\ S\ \{P \wedge \neg B\}}$

(2) $\dfrac{\{P\}\ S\ \{Q\},\ Q \wedge \neg B \supset Q}{\{P\}\ \textbf{repeat}\ S\ \textbf{until}\ B\ \{Q \wedge B\}}$

The first rule of consequence states that if the execution of a program S ensures the truth of the assertion R then it also ensures the truth of any assertion implied by R. Finally, the assertion P in the first iteration rule is called a *loop invariant,* since it remains unchanged over the iterations of the loop body.

Consider now the proof of a (sequential) program. Given a formula $\{P\}\ S\ \{Q\}$, a proof of this may be constructed by using the rules of Table 6.1. For example, suppose we have a program that performs multiplication of two positive integers by repeated additions. Then, after adding the preconditions and postconditions, the program might be as follows:

$\{(x > 0) \wedge (y > 0)\}$ (6.3)
    begin
        z := 0;
        u := x;
        repeat
            z := z + y
            u := u − 1
        until u = 0
    end
$\{z = x * y\}$

In order to demonstrate the truth of (6.3) one must systematically apply the available axioms and proof rules. Such a proof could be expressed in the form of a *proof outline,* where the assertions are interleaved with the program statements at appropriate points. For the above example, a possible proof outline is shown in (6.4).

$$\{(x > 0) \wedge (y > 0)\} \tag{6.4}$$
$$\begin{aligned}
&\textbf{begin} \\
&\quad z := 0; \\
&\quad u := x; \\
&\quad \{(z + u * y = x * y) \wedge (u > 0)\} \\
&\quad \textbf{repeat} \\
&\quad\quad z := z + y; \\
&\quad\quad u := u - 1 \\
&\quad\quad \{(z + u * y = x * y) \wedge (u \geq 0)\} \\
&\quad \textbf{until } u = 0 \\
&\quad \{(z + u * y = x * y) \wedge (u = 0)\} \\
&\textbf{end} \\
&\{z = x * y\}
\end{aligned}$$

It can easily be shown, for example, by using the axiom of assignment and the rule of composition, that

$$\{x > 0 \wedge y > 0\} z := 0; u := x \{(z + u * y = x * y) \wedge (u > 0)\} \tag{6.5}$$

Furthermore, using the same rules, we can show that

$$\{(z + u * y = x * y) \wedge (u > 0)\} \tag{6.6}$$
$$z := z + y; u := u - 1$$
$$\{(z + u * y = x * y) \wedge (u \geq 0)\}$$

Clearly

$$(z + u * y = x * y) \wedge (u \geq 0) \wedge \neg(u = 0) \tag{6.7}$$
$$\supset (z + u * y = x * y) \wedge (u > 0)$$

On application of the **repeat-until** rule to (6.6) and (6.7), it follows that

$$\{(z + u * y = x * y) \wedge (u > 0)\} \tag{6.8}$$
$$\quad \textbf{repeat } z := z + y;$$
$$\quad\quad u := u - 1$$
$$\quad \textbf{until } u = 0$$
$$\{(z + u * y = x * y) \wedge (u \geq 0) \wedge (u = 0)\}$$

The desired assertion $z = x * y$ follows directly from the postcondition in (6.8).

Since our present concern is the verification of S*A descriptions, proof rules and axioms must be constructed for S*A. These are given in Appendix A, which defines the complete language. For our present purpose it is sufficient for the reader to understand that the proof rules for the assignment statement and sequential control constructs in S*A are mostly identical to those given in Table 6.1. Additional rules required for parallel statements and other constructs specific to S*A will be introduced in the text as and when necessary.

6.5 DERIVATION OF A CORRECT MECHANISM

Returning to the Wilkes (1973) scheme, consider now the procedural part of the mechanism. This will be activated by a call to one of its public procedures from some point in the instruction interpretation cycle, after the opcode has been placed in csar. The mechanism will be deactivated at a point at which the microinstruction register contains the first microinstruction of the routine that interprets this code.[1]

In order to gain the greatest benefit from the use of a symbolic architectural language, we shall derive the design in such a way that it can be formally verified relative to some given set of specifications.

At the time of activation, the following preconditions will hold:

[PRE1] (csar.hiorder $= 0$) & (csar.opcode $= op_0 \ni 0 \leq op_0 \leq 2^{a+1} - 1$)
[PRE2] isn_reg $= I \ni 0 \leq I \leq 2^{i+1} - 1$
[PRE3] $(\exists k)(1 \leq k \leq 2^{i+1} - 1$ & isn_table[k].isn $=$ isn_reg)
[PRE4] $\forall j \in 1..2^{i+1} - 1)$
   (mm[isn_table[j].st_addr] $=$ isn_table[j].unique_id)

The preconditions [PRE1] and [PRE2] simply state that csar and isn_reg will have values defined within the specified ranges. The precondition [PRE3] signifies that there exists, at the time the mechanism is activated, an entry in isn_table corresponding to the value in isn_reg. The precondition [PRE4] states that the starting address of a microcode table as specified by an isn_table entry contains the same unique_id as specified in the "unique_id" field of the entry (cf. Fig. 6.5).[2]

---

[1] Note that since instruction interpretation is done in the proposed machine by firmware, this implies that the microprogram for instruction fetch and execute will be interrupted just after the fetch phase by the hardware mechanism about to be described. Of course, at the S*A level of description, this implementation detail will be invisible.

[2] From Fig. 6.2 we see that neither the word length of main memory nor the bit length of the **seq** unique_id have been specified. However, in order for [PRE4] to hold, and as Fig. 6.5 suggests, the unique_id at the head of each microcode table in main memory must occupy an entire word.

Let us assume that the opcodes are encoded such that if the opcode belongs to the general instruction set then csar.opcode[a] = 1; otherwise csar.opcode[a] = 0. Thus our first postcondition can be established as illustrated below.

[PRE5] csar.opcode[a] = 1
[POST1] Given [PRE1]–[PRE5] :
    mir.microinst = CS1 [csar.opcode[a−1 .. 0]].microinst &
    mir. next_addr = CS1 (csar.opcode[a−1 .. 0]].next_addr

That is, if the opcode is from the general set, then the microinstruction register must contain the corresponding first microinstruction from CS1.

On the other hand, if csar.opcode is from one of the special instruction sets, then we have either of the following cases A or B:

Case A. If there exists a valid microword in CS2 at location csar.opcode whose ISN matches that in isn_reg then (1) CS1 and CS2 should both remain unchanged, and (2) the microinstruction register should hold the value in CS2[csar.opcode].
This case is more formally expressed by the assertions:

[PRE6] csar.opcode[a] = 0 & CS2[csar.opcode].isn = isn_reg &
    CS2[csar.opcode].validity = 1
[POST2] Given [PRE1] − [PRE4] & [PRE6] : mir = CS2 [csar.opcode]

Case B. If the microword in CS2[csar.opcode] is invalid or its ISN does not match the value in isn_reg then (1) CS2[csar.opcode] must contain the first microinstruction from the proper microcode table; (2) the microinstruction register must also contain this newly loaded microinstruction; (3) CS1 must contain the remaining microinstructions of the same microroutine.

We decide that (3) above is to be realized by a call to a procedure called MICROLOADER—which, from our informal architecture, we know will be implemented in firmware. Providing MICROLOADER is not to be activated from anywhere else, and will not execute in parallel to the procedure we are presently constructing, it can be described as a private procedure in CHECK_CTLSTORES. Thus, we have partly resolved our modularization problem at this point.

Case B(1) can be precisely stated as:

[PRE7] csar.opcode[a] = 0 & (CS2[csar.opcode].isn ≠ isn_reg
   or CS2[csar.opcode].validity = 0)

[POST3] Given [PRE1 − PRE4] & [PRE7]:
    CS2 [csar.opcode].type_bit = curr_mcode_table.routine_part[k].
     type_bit

```
      & CS2 [csar.opcode].microinst = curr_mcode_table.routine_part[k].
        microinst
      & CS2 [csar.opcode].next_addr = curr_mcode_table.routine_part[k].
        next_addr
      & curr_mcode_table.index_part[k].opcode = csar.opcode
      & CS2 [csar.opcode].validity = 1
      & CS2 [csar.opcode].isn = isn_reg
```

Case B(2) is simply described by the assertion

[POST4] mir = CS2[csar.opcode].

We will not formalize the assertion corresponding to B(3), since its precise form will depend on exactly which part of CSI is to be loaded with the rest of the microroutine. In fact, this assertion will be part of MICROLOADER's postcondition. The only point to remember is that since MICROLOADER will require the microinstruction register for its execution, the action leading to [POST4] cannot be executed until MICROLOADER has completed its task.

For future reference I shall label the six clauses of [POST3] as [POST31]–[POST36]. Based on the postconditions [POST1]–[POST3] we obtain the skeletal outline of the main procedure shown in Fig. 6.6. Notice that the form of this procedure follows directly from the assertions [POST1]–[POST3].

We now need to construct STMT1 such that the assertion

$$\{[PRE1-PRE4] \ \& \ [PRE5]\} \ STMT1 \ \{[POST1]\} \qquad (6.9)$$

is satisfied. We can do this by means of the parallel statement:[3]

do mir.microinst := CS1[csar.opcode[a−1..0]].microinst □ mir. $\qquad$ (6.10)
next_addr := CS1[csar.opcode[a−1..0]].next_addr **od**

Similarly for STMT21, to make true the assertion

$$\{[PRE1-PRE4] \ \& \ [PRE6]\} \ STMT21 \ \{[POST2]\} \qquad (6.11)$$

---

[3] In the preceding section we considered the proof rule for sequential composition. While the general case of parallel composition will be discussed in Chapter 9, it is necessary to provide here the proof rule for the restrictive case of the parallel statement **do** S_1 □ S_2 □ . . . □ S_n **od** in which the statements $S_1, \ldots, S_n$ taken pairwise reference disjoint sets of variables. Let I_i, O_i denote respectively the input and output sets of variables for S_i. Then S_i, S_j ($i \neq j$) are said to reference disjoint sets of variables if $I_i \cap O_j = \emptyset \wedge O_i \cap I_j = \emptyset \wedge O_i \cap O_j = \emptyset$ where $\emptyset$ is the empty set. The proof rule for this special case is then simply:

$$\frac{\{P_1\} \ S_1 \ \{Q_1\}, \ \{P_2\} \ S_2 \ \{Q_2\}, \ \ldots, \ \{P_n\} \ S_n \ \{Q_n\}}{\{P_1 \wedge P_2 \wedge \ldots P_n\} \ \textbf{do} \ S_1 \ \square \ S_2 \ \square \ \ldots \ \square \ S_n \ \textbf{od} \{Q_1 \wedge Q_2 \wedge \ldots \wedge Q_n\}}$$

```
            {[PRE1 - PRE4]}
            if csar.opcode[a] = 1
         => {[PRE1 - PRE4] & [PRE₅]}
               /* implies general instruction set; so read microinstruction from *
               * CS1 */
               STMT1
               {[POST₁]}

            ||csar.opcode [a] = 0
         => /* implies special instruction set */
            {[PRE1 - PRE4] & ¬[PRE5]}
            if CS2 [csar.opcode].isn = isn_reg ∧
               CS2 [csar.opcode].validity = 1
            => /* microinstructions are already in CS1 and CS2 */
                  {[PRE1 - PRE4] & [PRE₆]}
                       STMT21
                  {[POST2]}

               else {[PRE1 - PRE4] & [PRE7]}
                   /* microroutine not in control stores */
                       STMT22 ;
                   {[POST3] & [POST4]}
      fi
   fi
```

Figure 6.6 Proof outline for main procedure in CHECK_CTL_STORE.

we may suggest

do mir.microinst := CS2[csar.opcode[a − 1..0]].microinst   (6.12)
□ mir.next_addr := CS2[csar.opcode[a − 1..0]].next_addr **od**

These statements are suggested quite naturally from the given preconditions and postconditions, and the semantics of the assignment and parallel statements in S*A (cf. also Sections 8.1 and 8.2 of Appendix A). It is a trivial task to show that (6.10) and (6.12) satisfy the assertions (6.9) and (6.11), respectively.

Remark. The basic approach to the derivation of the description will now be evident. We first construct assertions that the machine state must satisfy. The assertions specify relationships that must hold prior to the execution of a statement S (the precondition P) and on termination of S (the postcondition Q). From an examination of P and Q, S is then developed. It may then be verified that, given P, the execution of S leads to Q.

Thus, the development of an S*A description is guided by the assertions. We owe this approach to the work of E. W. Dijkstra (1976), D. Gries (1976a), and others on formal program derivation.

Consider now the assertion

$$\{[\text{PRE1-PRE4}] \ \& \ [\text{PRE7}]\} \ \text{STMT22} \ \{[\text{POST3}]\} \qquad (6.13)$$

Reflecting on [POST3] we note that what must first be determined is the value of the index k such that the first four clauses of [POST3] are satisfied. Furthermore, it is clear that since csar_opcode is not to be altered from its precondition [PRE1] value, k has to be found by some statement such that the fourth clause

[POST34]: curr_mcode_table.index_part[k].opcode = csar.opcode

is true. Thus we see that of the seven clauses in [POST3] and [POST4]:

1. [POST34] must first be satisfied so as to access the appropriate (kth) entry of curr_mcode_table.index_part.
2. [POST32], [POST33], [POST35], and [POST36] should precede [POST4] since the microinstruction register must be loaded with the new "first" microinstruction from CS2 after CS2 has been updated.
3. [POST31]-[POST33], [POST35], and [POST36] could well be realized by parallel statements if parallel paths and units appear feasible.

Thus, the assertion (6.13) can be *decomposed* to:

$$\begin{aligned}&\{[\text{PRE1-PRE4}] \ \& \ [\text{PRE7}]\} \\ &\quad \text{STMT 222} \\ &\{[\text{POST34}]\} \\ &\quad \text{STMT 221} \ \square \ \text{STMT 223} \ \square \ \text{STMT 224} \\ &\{[\text{POST31-POST33}] \ \& \ [\text{POST35}] \ \& \ [\text{POST36}]\} \\ &\quad \text{STMT 225} \ \{[\text{POST3} \ \& \ [\text{POST4}]\} \end{aligned} \qquad (6.14)$$

Consider now STMT 222. Since curr_mcode_table.index_part is an array-type data structure in main memory, we must search it iteratively.

Let "index" be a (private) variable that will be used to hold at any time the main memory address of some entry in curr_mcode.table.index_part:

privar index : **seq[..]bit**

The initial condition that index must satisfy is:

$$\text{index} = \text{address of curr\_mcode\_table.index\_part}[1] \qquad (6.15)$$

The search through the index part will terminate with the following conditions:

curr_mcode_table[index].opcode = csar.opcode **or** (6.16)
index = address of curr_mcode_table.routine_part[1]

Thus, we suggest as the *loop invariant*:[4]

[P]: address of curr_mcode_table.index_part[1] ≤ index ≤
    address of curr_mcode_table.routine_part[1]

Note that the initial condition (6.15) implies that [P] holds. We also define on the basis of the assertion (6.16) the following Boolean condition:

[B]: curr_mcode_table[index].opcode ⌉ = csar.opcode **&**
    index < address of curr_mcode_table.routine_part[1]

Then, we need to construct a statement STMT 222′ such that the assertion

$$\frac{\{P \wedge B\} \text{ STMT 222}' \{P\}}{\{P\} \textbf{ while } B \textbf{ do } \text{STMT 222}' \{P \wedge \neg B\}} \quad (6.17)$$

is satisfied. The assertion (6.17) is simply the proof rule for the **while** statement (cf. Table 6.1 and Appendix A, Section 8.2.6).

Before we can continue further with the refinement, we must define the entities "address of curr_mcode_table.index_part[1]" and "address of curr_mcode_table.routine_part[1]" in terms of declared data objects and synonyms. These may easily be extracted from Figs. 6.3–6.5. Thus,

$$mm[mcode\_table\_base + 1] \quad (6.18)$$

gives us the first address, since index_part starts immediately after the unique_id field of each microcode table. The second address is obtained by observing that the st_addr field in each index_part entry specifies the relative starting address of the corresponding routine_part. Thus, the absolute address of curr_mcode_table.routine_part[1] is given by:

mm[mcode_table_base + curr_mcode_table.index_part[1].st_addr]   (6.19)

In order to satisfy the initial condition (6.15), then, the value of the expression (6.18) must be assigned to index. And since, according to (6.17), index

---

[4] A loop invariant or an invariant property associated with a repetition statement (see Fig. 6.7) is a property P such that if P is true of the machine state initially (i.e., before entering the loop body), P will be true of the machine state after each traversal of the loop.

DERIVATION OF A CORRECT MECHANISM

must be compared in each iteration of the **while** statement, with the value of expression (6.19), we postulate a new private variable

privar st_addr_routine_pt : seq[..]bit

Before entering the loop, we assign the value of (6.19) to this data object. Thus STMT 222 will be refined to the following sequence:

```
/* place starting addresses of index_part and routine_part *                    (6.20)
 * in index and st_addr_routine_pt respectively */
index := mm[mcode_table_base + 1];
st_addr_routine_pt := mm[mcode_table_base +
       curr_mcode_table.index_part[1].st_addr];
{P}
/* search index_part for entry corresponding to
           csar. opcode */
while curr_mcode_table[index].opcode ¬ = csar.opcode ∧
   index < st_addr_routine_pt do index := index + 1 od
{P ∧ ¬B}
```

The assertion ¬B, which holds when the loop terminates, implies either of the following conditions:

[Q1] index < st_addr_routine_pt &
    curr_mcode_table[index].opcode = csar.opcode
[Q2] index = st_addr_routine_pt & [(∀index)
    (address of curr_mcode_table.index_part[1] ≤ index <
    address of curr_mcode_table.routine_part[1])
    (curr_mcode_table[index]. opcode ¬= csar.opcode)]

Clearly [Q2] indicates that there exists no entry in curr_mcode_table.index_part corresponding to the opcode in csar. This we define as an error condition that can be trapped out with:

```
{Q1 or Q2}                                                                       (6.21)
if index = st_addr_routine_pt => trap INVALID_OPCODE fi
{Q1}
```

where INVALID_OPCODE is a private error handling procedure. Since the assertion [Q1] implies [POST34], this completes the construction of STMT 222 of (6.14). It consists of the statement sequence (6.20); (6.21).

The remaining statements of (6.14) are still to be developed. Since [POST31]–[POST33], [POST35], and [POST36] can all be realized by assignment statements that are pairwise disjoint in their use of data objects we can use a parallel statement:

do CS2 [csar.opcode].type_bit :=
    curr_mcode_table[curr_mcode_table[index].st_addr].
                                                    type_bit
  ☐ CS2 [csar.opcode].microinst :=
    curr_mcode_table[curr_mcode_table[index].st_addr].
                                                    microinst
  ☐ CS2 [csar.opcode].next_addr :=
    curr_mcode_table[curr_mcode_table[index].st_addr].
                                                    next_addr
  ☐ CS2 [csar.opcode].validity := 1
  ☐ CS2 [csar.opcode].isn := isn_reg
od

Finally, STMT 225 (in 6.14), such that its postcondition is

$$[POST4]: mir = CS2[csar.opcode]$$

will simply be

$$mir := CS2[csar\_opcode].$$

We now have all the components of STMT 22. Only a single mechanism was required, and this is shown in its final form as Fig. 6.7. The mechanism includes two other private procedures whose details are not shown here.

Remarks

1. It can be seen that the formal description makes it considerably easier to test whether or not form matches requirements. Consider the requirement (R6.1): if the opcode belongs to the general instruction set, CHECK_CTLSTORE.MAIN indicates that the appropriate microinstruction is located in CS1 and loaded into mir in the usual manner. The only additional time spent is in the testing of csar.opcode[a]. Likewise, in the case of (R6.2), if the opcode is from a special instruction set for which the microroutine is already in control store, the added overhead lies in testing for the condition

   CS2[csar.opcode].isn = isn_reg & CS2[csar.opcode].validity = 1

   We cannot comment on whether (R6.3) holds or not since this will depend on the internal design of MICROLOADER. Finally—and this is in fact apparent from the informal design—the separation of microroutines according to instruction sets limits the extent to which microcode can be shared between instructions.

2. Every design has an esthetic content. By this I mean not necessarily a visual attribute (though in some obvious contexts, notably building architecture, this may indeed dominate), but an internal characteristic that originates in whatever principles underlie the design. In the case of particularly complex systems—a bridge, a building, a natural form, or even a proof of some theorem—the esthetics is (paradoxically) of a technical nature, in that its significance and context can be perceived only if the beholder has the necessary technical knowledge. A structure, for example, that appears commonplace to the eye may have been designed on a new principle of engineering science or structural theory. Its esthetics may then lie in the manner in which the principle has been used in this particular design. A house may have been designed to maximize the use of available solar energy. Its esthetics may lie in how the architect's conception embodies this objective.

The esthetics of computer architecture has probably never been explicitly addressed, perhaps because the issue may seem irrelevant in computer design. Yet the mathematician and the physicist have often embraced a proof or a theory that, in addition to being correct or plausible, as the case may be, is "beautiful," "elegant," or "breathtaking in its simplicity." William of Occam's principle embodies the notion of simplicity—a powerful esthetic factor.[5]

Thus, esthetics has an honorable place as a design criterion in computer science in general and computer architecture in particular.[6]

In the case of exoarchitecture, which we shall consider in more detail in a later chapter, this will essentially be an esthetics of functionalism. In the case of endoarchitecture the esthetics will be determined in part by how solutions are obtained to certain technical problems (the algorithm used), but also in considerable part by the clarity of the final form, the logical principles employed, the economy of the form, and the efficiency of the design. Finally, its esthetics will also be a function of the extent to which it absorbs new results in theory and technology.

It is my belief that without a formal architecture as I have proposed here, one cannot adequately assess the esthetic content of an architecture design. Let us also note in passing that the design depends not only on the language used, but also on how the architect uses the language.

---

[5] The principle, due to the fourteenth century English philosopher William of Occam and known as Occam's Razor, states that the hypothesis or explanation involving the fewest assumptions is the best.
[6] "When we recognize the battle against chaos, mess, and unmastered complexity as one of computing science's major callings we must admit that 'Beauty is our Business.' " (E. W. Dijkstra, "Some Beautiful Arguments Using Mathematical Induction," *Acta Informatica*, Vol. 13, Fasc. 1, Jan. 1980.)

```
mech CHECK_CTL_STORE;
/* definition of state variables */
glovar csar: tuple hiorder : seq[b..a+1]bit; /* CS1 addr reg is the whole of */
              opcode  : seq[a..0]bit  /* csar; CS2 add.reg is csar. */
         endtup                        /* opcode */
glovar isn_reg : seq[i..0]bit         /* curr inst set no register */
glovar isn_table : array[1..2^{i+1}] of /* table of inst.set.numbers */
                  tuple isn : seq[i..0]bit;
                        unique_id : seq[..]bit;
                        st_addr : seq[..]bit; /* st.addr. of microcode table */
                                              /* for this isn */
                        length : seq[..]bit;  /* length of microcode table
                                                 for this isn */
                  endtup
glovar mm ; array[..] of seq[..]bit;    /* main memory */
const one : bin(1) 1;

type sh_mword = tuple typebit : bit;           /* short microword format */
                      microinst : seq[..]bit;  /* microinstruction */
                      next_addr : seq[b..0]bit
                endtup;
type lg_mword = tuple validity:bit;            /* long microword format */
                      isn : seq[i..0]bit;
                      typebit : bit;
                      microinst : seq[..]bit   /* microinstruction */
                      next_addr : seq[b..0]bit
                endtup;
glovar mir : lg_mword;                         /* microinstruction register */
privar CS1 : array[0..2^{b+1}-1] of sh_mword with csar, csar.opcode;
privar CS2 : array[0..2^{a+1}-1] of lg_mword with csar.opcode;
type mroutine = array[.] of sh_mword;
type mcode_table = tuple unique_id : seq[..]bit;
                         index_part : array [ ] of
                              tuple opcode : seq[a..0] bit;
                                    st_addr : seq[..]bit;
                                    length : seq [ ] bit
                              endtup;
                         routine_part : array [ ] of mroutine;
                         table_end : bin (..) 11..1
                   endtup;
syn mcode_table_base = isn_table[isn_reg].st_addr;
syn mcode_table_limit = isn_table[isn_reg].length;
syn curr_mcode_table : mcode_table =
     mm[mcode_table_base..mcode_table_base+mcode_table_limit-1];
privar st_addr_routine_pr, index : seq[..]bit;
```

Figure 6.7 Detailed description of CHECK_CTL_STORE.

```
proc MAIN;
   /* decode opcode */
   if csar.opcode[a] = 1 =>
      /* implies reg.inst.set; read microinst.from CS1 */
      do mir.minst := CS1 [csar.opcode[a-1..0]].minst
       □ mir.next_addr := CS1 [csar.opcode[a-1..0]].next_addr
      od
   || csar.opcode[a] = 0 =>
      /* implies special inst.set */
      if CS2  csar.opcode .isn = isn_reg ∧ CS2[csar.opcode].validity = 1
         /* microroutine for sp.inst set already loaded in CS2 and CS1 */
         => do mir.minst := CS2 [csar.opcode[a-1..0]].minst
               mir.next_addr := CS2 [csar.opcode[a-1..0]].next_addr
            od
      || else =
         /* microroutine not in CS2/CS1; place st_addr of index_part in index */
         index := mm[mcode_table_base + 1];
         /* place st.addr of routine part in st_addr.routine_pt */
      st_addr_routine_pt :=
         mm[mcode_table_base + curr_mcode_table.index_part[1].st_addr]
         /* search index_part for entry corresp. to csar.opcode */
         while curr_mcode_table[index].opcode ¬ = csar.opcode ∧
            index < st_addr_routine_pt do index := index + 1 od;
         /* if entry corresp. to csar/opcode not in index_part then trap */
         if index = st_addr_routine_pt => trap INVALID_OPCODE fi;
         /* otherwise load first microinstruction into CS2 */
         /* load the whole microword from CS2 into mir */
         do CS2[csar.opcode].typebit :=
               curr_mcode_table[curr_mcode_table[index].st_addr].type_bit
          □ CS2[csar.opcode].minst :=
               curr_mcode_table[curr_mcode_table[index].st_addr].minst
          □ CS2[csar.opcode].next_addr :=
               curr_mcode_table[curr_mcode_table[index].st_addr].next_addr
          □ CS2[csar.opcode].validity := 1
          □ CS2[csar.opcode].isn := isn_reg
         od;
         call MICROLOADER
         mir := CS2[csar_opcode];
      fi
   fi
endproc
proc INVALID_OPCODE priv; ..... endproc
proc MICROLOADER priv;    ..... endproc
endmech
```

Figure 6.7 (*Continued*)

6.6 BIBLIOGRAPHIC NOTES

The microcode loading problem is based on Wilkes (1973). For additional discussion of this same problem the reader is referred to Manville's (1973) thesis.

The seminal works on program correctness are the classic papers by Floyd (1967) and Hoare (1969). Several texts have been devoted to program correctness, notably Alagic and Arbib (1978), Jones (1980), Manna (1974), Dijkstra (1976), de Bakker (1980), and Gries (1982). The first and last of these references are specially recommended for their treatment of the so-called axiomatic (Floyd-Hoare) method. The notion of formal program derivation using assertions to guide the process has been advocated by Dijkstra (1976) and Gries (1976a, b, 1982), among others.

SEVEN

AN ASYNCHRONOUS ARCHITECTURAL SYSTEM

7.1 OVERVIEW OF THE SYSTEM

In this chapter we shall consider an example of an asynchronous architectural system as it may be described in an architectural language.

For this purpose the system selected is based loosely on the architecture of the MU5 instruction buffer unit (IBU). It must be pointed out, however, that this description should not be construed as representing the instruction buffer unit as it actually exists in the MU5. My understanding of this system is based on the monograph by Morris and Ibbett (1979); because this source provides an informal treatment of the architecture, some of the problems characteristic of the informal design process were found to be present. Consequently, any attempt to describe exactly and formally the actual IBU subsystem becomes somewhat futile; what is given below must, instead, be viewed as a description of an architectural system cast in the *spirit* of the MU5 architecture. For convenience, however, I shall employ the same names for the different functional components as are used in the MU5 literature.

The IBU consists of two principal components: DATAFLOW and STORE_REQ. Our concern here is with the former. The IBU as a whole interacts on the one hand with the "store access control" (SAC) system and on the other with the "primary operand unit" (PROP). The SAC system sends 128-bit instruction packets (eight instructions/packet) to DATAFLOW, and it is the latter's responsibility to accept each such packet and place it in an input buffer. The DATAFLOW system in turn sends out 16-bit instructions to PROP from an output buffer.

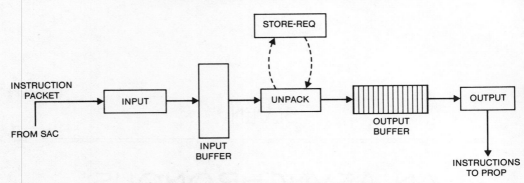

Figure 7.1 Instruction pipeline: flow structure.

Thus DATAFLOW inputs packets from SAC, unpacks instructions from the input packet, lining them up in an output buffer, and outputs instructions to PROP. While unpacking, DATAFLOW must perform certain other functions, outlined below.

The DATAFLOW system thus serves to prepare instructions for the later phases of operand access and fetch and to supply a steady flow of instructions at the maximum possible rate to PROP. For the sake of maintaining the highest possible throughput through the system we desire that the input, unpack, and output functions be done concurrently, independently, and asynchronously. Thus the task at hand is to describe DATAFLOW as a single asynchronous system whose overall effect is to establish a pipeline through which instructions pass.

Figure 7.1 shows this pipeline diagrammatically, indicating the necessary registers, the processes that control the pipeline, and the nature of the dataflow through it. Note that this diagram is not intended to show the actual data paths in DATAFLOW, nor does it depict by its constituent boxes actual hardware units. Instead, INPUT, UNPACK, and OUTPUT denote control logic units that are logically distinct but in actual realization may possibly be embedded in the same hardware module. Thus, Fig. 7.1 depicts in a highly schematic way the intended flow of information through the storage devices and control logic of DATAFLOW. It is, in fact, a crude but useful representation of its *flow structure,* in which much of the important information has been totally hidden from view. Diagrammatic flow structures, together with natural language descriptions and diagrammatic path structures (i.e., conventional block diagrams), constitute an informal description of an architecture.

7.2 FORMAL DESCRIPTION

As already mentioned, the purpose of INPUT is to accept a 128-bit packet of eight instructions and place it in the input buffer. There are two conditions that have to be satisfied before INPUT is able to do this:

1. A packet must have been sent by the sender, namely, SAC. For this purpose we may define a synchronizer "packet_sent" and a channel "bus_A" to be shared between SAC and INPUT:

    sync packet_sent : **bit** (0) ; /* 0 implies packet not sent */
    chan bus_A : seq[127..0]bit

2. The input buffer must have been completely emptied (by UNPACK). To denote this condition we define another synchronizer:

    sync ibuffer_empty : **bit** (1) ; /* 1 implies empty */

Thus, the rate at which INPUT can fill the input buffer is determined almost entirely by the rate at which SAC can deliver instructions and the rate at which UNPACK can empty the input buffer. If SAC delivers one 128-bit packet every n nanoseconds, it is desirable that UNPACK empty the input buffer at roughly the same rate.

At the endoarchitectural level, as described in S*A, it is not possible (nor it is intended) to embed quantitative performance characteristics in the formal description/design. We may merely indicate implicitly the qualitative characteristics of performance by the form of the formal design.

Quantitative performance characteristics, and their influence on the design, may enter the design process in essentially two ways:

1. By conducting simulation studies of the architecture as specified in S*A, introducing appropriate performance parameters during simulation, and possibly modifying the design as a result of these studies.
2. By introducing timing characteristics in the refinement of the architecture to lower levels of abstraction (e.g., the register-transfer or microarchitectural levels). In this case, any modification to the design would (ideally) be restricted to the lower level and be invisible at the level of endoarchitecture.

Returning to the problem at hand, when INPUT has filled its input buffer, it must communicate this fact to UNPACK. The synchronizer "ibuffer_full" is defined for this purpose:

sync ibuffer_full : **bit** (0) ; /* 0 implies input buffer empty */

Clearly, given that INPUT, UNPACK, and OUTPUT must work concurrently, they must be placed in separate mechanisms. The complete **mech** for INPUT can thus be described as shown in Fig. 7.2. The global variable "input_buffer" is to be shared between INPUT and UNPACK. We ensure that the two corresponding public procedures INPUT.IBUFFER_FILL and UNPACK.BUFFER_TRANSF (see Fig. 7.3) do not access "input_buffer"

```
mech INPUT;
    type inst = seq [15..0] bit;
    glovar input_buffer : array [0..7] of inst;
    sync packet_sent : bit (0);
                    /* 0 implies packet not sent */
    sync ibuffer_full : bit (0);
                    /* 0 implies input_buffer empty */
    sync ibuffer_empty : bit (1);
                    /* 1 implies input_buffer empty */
    sync bus_A_empty : bit (1);
                    /* 1 implies bus_A empty */
    chan bus_A : seq [127..0] bit;

    proc IBUFFER_FILL;
        forever do
          /* await till input_buffer has been emptied
                              by UNPACK */
            await ibuffer_empty
                do /* accept input packet if available */
                    await packet_sent
                        do input_buffer := bus_A;
                            sig bus_A_empty
                        od;
                    sig ibuffer_full
                od
        od
    endproc
endmech
```
Figure 7.2 Instruction pipeline: input.

FORMAL DESCRIPTION

```
mech UNPACK;
  type inst = seq[15..0]bit;
  chan jump_trace_resp_line: bit;
  glovar input_buffer : array[0..7]of inst;
  glovar output_buffer : array[0..7] of
                          tuple instrn : inst;
                                lis    : bit(1); /* 1 ⊃ last inst. in seq */
                                full   : bit(0)  /* 1 ⊃ full */
                          endtup;
  privar ibuffer_pointer : seq[3..0]bit;
  sync ibuffer_full  : bit(0); /* 0 ⊃ empty */
  sync ibuffer_empty : bit(1); /* 1 ⊃ empty */
  sync obuffer_top_empty : bit(1); /* 1 ⊃ empty */
  sync store_req_jump_store_start : bit(0);
  sync is_branch : bit(0);

  proc BUFFER_TRANSF;
    forever do
      /* wait till input buffer full before unpacking */
        await ibuffer_full do
          ibuffer_pointer := 0;
          repeat /* wait till top of output buffer is emptied (by "output") */
            await obuffer_top_empty
            do output_buffer[7].instrn := input_buffer[ibuffer_pointer]
              □ output_buffer[7].full := 1;
                sig store_req_jump_store_start;
                await is_branch
                  do output_buffer[7].lis := jump_trace_resp_line od;
                if output_buffer[7].lis = 0 =>
                               ibuffer_pointer := ibuffer_pointer + 1 fi
            od;
            sig obuffer_top_full;
          until ibuffer_pointer = 8 ∨ output_buffer[7].lis = 1
          /* ibuffer_pointer = 8 means all 8 instructions transferred */
          /* successfully from input_buffer to output_buffer */
          /* signal to "input" to fill input buffer */
          sig ibuffer_empty
        od
    od /* end of forever loop */
  endproc
endmech
```

Figure 7.3 Instruction pipeline: unpack.

simultaneously by means of the synchronizer pair "ibuffer_empty" and "ibuffer_full." Note also that the procedure INPUT.IBUFFER_FILL, once initiated, remains active forever.

The UNPACK mechanism performs the following:

1. It transfers 16-bit instructions one at a time from "input_buffer" to "output_buffer" and lines them up.
2. Just after transferring an instruction to the top (end) of the "output_buffer" line it issues a signal to a procedure (call it JUMP_TRACE, using the MU5 term) in the STORE_REQ mechanism and waits for a (binary-valued) response from the latter that indicates whether or not there is an entry in the STORE_REQ associative store for this instruction.

```
mech OUTPUT;
   glovar output_buffer : array[0..7] of
                             tuple instr : inst;
                                   lis   : bit(1); /* 1 implies last inst in seq */
                                   full  : bit(0)  /* 1 implies full */
                             endtup
   privar obuffer_ptr : seq[2..0] bit;
   sync obuffer_top_full : bit(0);    /* 1 ⊃ full */
   sync obuffer_top_empty : bit(1);   /* 1 ⊃ empty */
   sync busB_empty : bit(1);          /* 1 ⊃ busB empty */
   sync busB_full  : bit(0);          /* 0 ⊃ busB empty */
   chan busB : seq[15..0]bit

   proc OBUFFER_EMPTY;
      forever do
         if output_buffer[0].full => await busB_empty
                                       do busB := output_buffer[0].instrn;
                                          sig busB_full
                                       od fi;
            do output_buffer[0].full := 0 □ obuffer_ptr := 1 od;
            repeat
               output_buffer[obuffer_ptr-1] := output_buffer[obuffer_ptr]
               output_buffer[obuffer_ptr].full := 0;
               obuffer_ptr := obuffer_ptr +1
            until obuffer_ptr = 7;
            await obuffer_top_full do output_buffer[6] := output_buffer[7];
                                      sig obuffer_top_empty
                                   od
      od
   endproc
endmech
```

Figure 7.4 Instruction pipeline: output.

3. If the response indicates that there is an entry in the associative store then the instruction at the top of "output_buffer" is a branch for which the "jump to" address is immediately available in the associative store entry. The instruction is marked as "last instruction in sequence" (LIS) and the remaining instructions in "input_buffer" (which were in sequence to LIS) are "locked" out and thus prevented from being passed to "output_buffer." (The next sequence of instructions to be executed start at the "jump to" address and these must fill "input_buffer," overwriting the locked out instructions already there.)

The UNPACK mechanism is shown in Fig. 7.3.

Finally, we consider OUTPUT. This mechanism will transmit the contents of "output_buffer[0]" (the first element in "output_buffer") to PROP for the first stage of the instruction's execution. It will then transfer from "output_buffer[2]" to "output_buffer[1]," and so on. Clearly, OUTPUT must be synchronized with both UNPACK and PROP. A mechanism satisfying these conditions is described in Fig. 7.4.

All three mechanisms are defined as nonterminating. They are initially activated by an **init** statement as shown below:

system DATAFLOW ;
    mech INPUT ; .... ; **endmech** ;
    mech UNPACK ; .... ; **endmech** ;
    mech OUTPUT ; .... ; **endmech** ;

    init INPUT.IBUFFER_FILL,
            UNPACK.BUFFER_TRANSF,
                OUTPUT.OBUFFER_EMPTY ;
endsys

7.3 BIBLIOGRAPHIC NOTES

The most comprehensive and complete account of the MU5 computer system is the monograph by Morris and Ibbett (1979). A discussion of the system in its earliest stages of development is given in Kilburn et al. (1969), while Sumner (1974) provides a retrospective assessment. Ibbett (1972) is an excellent discussion of the MU5 instruction pipeline. A complete bibliography on the MU5 is given in Morris and Ibbett (1979).

The S*A description of the instruction buffer unit was first outlined (in part) in Dasgupta (1981) and later described more fully in Dasgupta (1982).

EIGHT

ON THE CORRECTNESS OF ASYNCHRONOUS ARCHITECTURAL SYSTEMS

8.1 INTRODUCTION

Verifying the design of asynchronous systems is considerably more complex than verifying sequential systems because of the added dimension of concurrency between mechanisms in the former. As in the case of parallel programs, when several hardware processes are active in parallel the results may depend on the generally unpredictable order in which actions from different processes are executed.

In this chapter I shall present an application of a technique proposed by Owicki and Gries (1976) to the verification of asynchronous architectural systems. The Owicki-Gries technique is basically intended for proving the correctness of parallel programs, but we shall see here that it is equally applicable to the domain of hardware architecture, provided the latter is described at an appropriate level of abstraction.

Specifically, we shall reconsider DATAFLOW, the system described in the previous chapter, in the light of correctness issues. Our objective is to demonstrate its *partial correctness* (this term was explained in Section 6.4) by using the Owicki-Gries technique.

8.2 PRINCIPLES OF THE OWICKI-GRIES TECHNIQUE

Owicki and Gries (1976) extended the Floyd-Hoare proof method (described earlier in Section 6.4) to the domain of parallel programs in the following way. The conventional Pascal-like sequential language is firstly augmented with two new constructs, one to initiate concurrent processes, the other to coordinate and synchronize such processes.

Let $S_1, S_2, \ldots, S_n$ be the statements. Then the statement

$$\textbf{cobegin } S_1 \parallel S_2 \parallel \ldots \parallel S_n \textbf{ coend}$$

causes $S_1, \ldots, S_n$ to be executed in parallel. Execution of this statement terminates only when execution of all the statements $S_1, \ldots, S_n$ have terminated. An important characteristic of the **cobegin** statement is that *no assumptions are made concerning the relative speeds of the individual processes $S_1, \ldots, S_n$.*

It is assumed that each assignment statement be executed, and each expression evaluated, as an indivisible action. However, we may relax this assumption if programs satisfy the convention below.

Convention 8.1. Each expression E may refer to at most one variable y if that can be changed by another process while E is being evaluated, and E may refer to y at most once. A similar restriction holds for the assignment statement $x := E$.

With this convention the only indivisible action need be the memory reference. Suppose statement S_i references variable x while $S_j (i \neq j)$ is altering its value. Then we require that the value of x obtained by S_i be the value of x either before or after the assignment to x. The indivisibility of memory reference assures us this property.

The second statement is of the form

await B then S

where B is a Boolean expression and S is a statement not containing a **cobegin** or another **await** statement. When a process attempts to execute an **await** it is delayed until the condition B is true. Then the statement S is executed as an indivisible action. If two or more processes are waiting for the same condition B, any one of them may be allowed to proceed when B becomes true, while the others continue to wait.

It is also stipulated that the evaluation of B is part of the indivisible action of the **await** statement. Another process may not change variables so as to make B false after B has been evaluated, but before S begins execution.

Formally, these constructs are defined as follows:

$$\frac{\{P \wedge B\} \; S \; \{Q\}}{\{P\} \; \textbf{await B then } S \; \{Q\}} \tag{8.1}$$

$$\frac{\{P_1\}\ S_1\ \{Q_1\},\ \ldots,\ \{P_n\}\ S_n\ \{Q_n\}\ \text{are interference-free}}{\{P_1 \wedge \ldots \wedge P_n\}\ \textbf{cobegin}\ S_1 \parallel \ldots \parallel S_n\ \textbf{coend}\ \{Q_1 \wedge \ldots \wedge Q_n\}} \quad (8.2)$$

The definition for the **cobegin** requires us to understand the meaning of the term *interference-free*. The simplest way of ensuring that a set of parallel processes do not interfere with each other is to have the processes operate upon disjoint sets of variables. This is clearly too restrictive. A more useful rule is to require that certain assertions used in the proof $\{P_i\}\ S_i\ \{Q_i\}$ of processes S_i are left true (i.e., are unaffected) under the parallel execution of the other processes. Thus, if these assertions are not falsified then the proof $\{P_i\}\ S_i\ \{Q_i\}$ will still hold, and hence Q_i will remain true upon termination. The invariance of an assertion P under the execution of a statement S is denoted by the formula

$$\{P \wedge \text{pre}(S)\}\ S\ \{P\}$$

We may now define *interference-free*:

Definition. Given a proof $\{P\}\ S\ \{Q\}$ and a statement T with precondition pre(T), we say that T does not interfere with $\{P\}\ S\ \{Q\}$ if the following conditions hold:

1. $\{Q \wedge \text{pre}(T)\}\ T\ \{Q\}$.
2. Let S' be any statement within S but not within an **await**. Then $\{\text{pre}(S') \wedge \text{pre}(T)\}\ T\ \{\text{pre}(S')\}$.

Intuitively, a statement T does not interfere with a proof $\{P\}\ S\ \{Q\}$ if T's execution has no effect on the truth of the precondition and the final postcondition used in the proof of S.

Definition. $\{P_1\}\ S_1\ \{Q_1\},\ \ldots,\ \{P_n\}\ S_n\ \{Q_n\}$ are interference-free if the following holds. Let T be an **await** or assignment statement (which does not appear in an **await**) of S_i. Then for all j, $j \neq i$, T does not interfere with $\{P_j\}\ S_j\ \{Q_j\}$.

In proving parallel programs it is often necessary to use *auxiliary variables*. These variables are required only for the proof of correctness and other properties, and not in the program itself, and typically record the state of execution of the program or part of the program. For example, in the well known producer-consumer problem—in which a producer process generates a stream of values for a consumer process—a buffer (array) may be used for the producer to deposit its values and the consumer to take previously generated values from. Mutual exclusion on buffer elements is ensured through the use of semaphores. A proof of correctness of the overall parallel program may require the use of auxiliary variables that keep track of the number of P and V operations on each semaphore. Thus, these variables may be added to the original program, and assignments made to them for purposes of correctness proofs.

Definition. Let AV be a set of variables that appear in S only in assignments $x := E$, where x is an AV. Then AV is an auxiliary variable set for S.

Auxiliary Variable Transformation. Let AV be an auxiliary variable set for S', and let P, Q be assertions that do not contain free variables from AV. Let S be obtained from S' by deleting all assignments to the variables in AV. Then

$$\frac{\{P\}\ S'\ \{Q\}}{\{P\}\ S\ \{Q\}}$$

With the help of the above definitions and concepts, we arrive at the essence of the Owicki-Gries technique. Given a parallel program expressed as a **cobegin** statement, we first consider each individual sequential process S_i and prove its correctness, ignoring parallel execution completely. We then show that the execution of each of the other processes does not interfere with the *proof* of S_i. We can then use (8.2) to infer the postcondition of the **cobegin** statement.

8.3 APPLICATION TO ARCHITECTURAL VERIFICATION

In applying this technique to architectural systems, we must first ascertain whether the Owicki-Gries model of parallel programs remains valid in the architectural context and, if this is not so, what kinds of transformations may legitimately be applied to architectural descriptions so that they do conform to the model.

In our case, since architectures are described in S*A, it is this language that must now be examined from this point of view. We shall therefore consider each of the crucial concepts in the Owicki-Gries technique and examine their implications for S*A.

8.3.1 Forms of Parallelism

In S*A, parallelism may be expressed in two ways and at two levels. Within a procedure, a set of concurrent actions may be specified in the form $S_1 \square \ldots \square S_n$, possibly separated from preceding and following sequential statements by the bracket pair **do** ... **od**. Any statement S_{n+1} following this parallel statement begins execution only when $S_1, \ldots, S_n$ have all terminated. This form obviously conforms to the **cobegin** statement in the Owicki-Gries model. Here, parallelism is confined to actions within the same procedure (and therefore within the same mechanism).

Secondly, two or more mechanisms may be active simultaneously, as a result of activation through **init** statements or through the invocation of a procedure in one mechanism by a procedure in another by means of the **act** statement. That parallelism exists in such a case is understood implicitly. Further, unlike the case of intraprocedure parallelism, the notion of a succeeding statement (or a procedure) beginning execution when the parallel procedures have terminated has no meaning in this context.

8.3.2 Synchronization

The S*A language has two forms of the **await** construct, **await** B and **await** x, where B is a boolean expression and x a synchronizer. Each may appear by itself, followed sequentially by a statement S (i.e., as **await** . . . ; S; . . .) or in the form

$$\textbf{await} \ldots \textbf{do} \text{ S } \textbf{od} \tag{8.3}$$

the distinction being that in the latter it is not possible to use a **goto** command to reach a statement in S from outside S. For convenience we shall, in this discussion, assume the more general form (8.3), since we can always transform a statement **await** B; S_1 (**await** x; S_1) to the equivalent statement **await** B **do od**; S_1 (**await** x **do od**; S_1), in which case S in (8.3) would simply be the empty statement.

The **await** B **do** S **od** construct in S*A is *almost* semantically identical to the Owicki-Gries **await** statement except that in the former S may contain embedded parallel or other **await** statements. The other form of the S*A **await** uses special variables called *synchronizers*. In executing this statement x is evaluated; if $x > 0$ it is decremented by 1 and S is then executed. The test and decrement is indivisible in that no other statement anywhere else in the system may reference or update x while this is going on. The statement S may contain here also other parallel and **await** statements.

8.3.3 Indivisibility of the Assignment Statement

In the Owicki-Gries model it is assumed that the assignment statement is either indivisible or that every expression or statement satisfies convention 8.1. The latter assumption appears undesirably restrictive, since it forbids such statements as $x := x + 1$ being executed in parallel with a statement such $y := x$. We shall therefore ignore this convention. This means that in order to use the Owicki-Gries method S*A descriptions must satisfy the following convention.

Convention 8.2. All expressions and assignment statements in an S*A description will be regarded as indivisible statements.

How reasonable is this convention? We must first note that if applied to parallel *programs* written in a high-level Pascal-like language, this would be wholly unrealistic, since each assignment statement is invariably translated into several machine instructions. Hence the introduction of Convention 8.1 by Owicki and Gries.

Our concern here is with the description of architectures. Whether an S*A statement realistically represents an indivisible action will depend on the fineness of the description or whether we may reasonably implement the statement as an indivisible action.

Example 8.1

In Fig. 7.2, the statement input_buffer: = bus_A may quite reasonably be regarded as indivisible, since its action may be implemented by a connection between bus_A and input_buffer such that the process of loading information from bus_A into input_buffer is, in fact, a single indivisible unit of action.

Example 8.2

Consider the statement in Fig. 7.3:

$$\text{ibuffer\_pointer} := \text{ibuffer\_pointer} + 1 \tag{8.4}$$

Whether this is indivisible or not will depend on how this action is to be realized; for example, if a counter is used to implement both the memory object "ibuffer_pointer" and the means of incrementing its value, then clearly the indivisibility of the statement can be assured. On the other hand, if a general purpose ALU is used for incrementing the contents of a register "ibuffer_pointer" it may be necessary to go through a sequence of more primitive assignments, for example:

do ail : = ibuffer_pointer □ air : = 1 **od**;
aout : = ail + air;
ibuffer_pointer : = aout

where "ail" and "air" are left and right inputs to the ALU respectively, and "aout" denotes the ALU output buffer. In this case the original S*A statement is not indivisible.

Thus, the application of the Owicki-Gries technique to the verification of architectural designs rests critically on the architectural description satisfying Convention 8.2. This also suggests that an appropriate criterion for the fineness of resolution of an architectural description is whether the individual assignment statements in it are realistically indivisible. Further, in order for a given proof to remain valid at lower levels of design (or implementation), a given S*A description would provide clues suggesting how the design may be cast in hardware (e.g., the use of a counter to implement the statement [8.4]).

8.4 VERIFICATION OF DATAFLOW

Keeping these caveats in mind, we shall now provide a proof of the correctness of DATAFLOW, the system described in the previous chapter. As suggested by Owicki and Gries, the technique involves two steps: we first prove the correctness of the individual procedures in each of the three mechanisms, ignoring the fact that they are being executed in parallel; we then show that the procedures are interference-free.

In order to use this techique we apply a common transformation to all the **await** statements appearing in DATAFLOW. Given the statement

$$\textbf{await } x \textbf{ do } S \textbf{ od}$$

where x is a **bit** type synchronizer, we first transform it to

$$\textbf{await } x \textbf{ ; } S$$

and then convert the new **await** to the equivalent S*A form

$$\textbf{await } x = 1 \textbf{ do } x := 0 \textbf{ od} \qquad (8.5)$$

Statement (8.5) is semantically identical to the **await** statement in the Owicki-Gries model.

Consider now the procedure INPUT_BUFFER_FILL in DATAFLOW (Fig. 7.2). The synchronizer pair "packet_sent" and "bus_A_empty" is used to coordinate information transfer between SAC (not shown here) and INPUT, while "ibuffer_full" and "ibuffer_empty" are similarly used by INPUT and UNPACK. We define the assertions

I_1 : (packet_sent = 1 $\wedge$ bus_A_empty = 0)
  $\vee$ (packet_sent = 0 $\wedge$ bus_A_empty = 1)
I_2 : (ibuffer_full = 0 $\wedge$ ibuffer_empty = 1)
  $\vee$ (ibuffer_full = 1 $\wedge$ ibuffer_empty = 0)

We then reproduce the procedure in Fig. 7.2 augmented by these and other assertions inserted appropriately, and with the **await** statements suitably transformed (Fig. 8.1). Note also that the two **sig** statements in this version have been specified as parallel actions.

```
proc IBUFFER_FILL
    forever do
        {I_1 ∧ I_2}

        await ibuffer_empty = 1 do ibuffer_empty := 0 od;
        {I_1 ∧ ibuffer_full = 0 ∧ ibuffer_empty = 0}
        await packet-sent = 1 do packet_sent = 0 od;


        {ibuffer_full = 0 ∧ ibuffer_empty = 0 ∧ packet_sent = 0 ∧ bus_A_empty = 0}

        input_buffer := bus_A;
        do sig bus_A_empty ☐ sig ibuffer_full od

        {(input_buffer = bus_A) ∧ (ibuffer_full = 1 ∧ ibuffer_empty = 0) ∧
            (packet_sent = 0 ∧ bus_A_empty = 1)}
        {I_1 ∧ I_2 ∧ input_buffer = bus_A}
    od
endproc
```

Figure 8.1 Proof outline for INPUT.IBUFFER_FILL.

Figure 8.1 in fact provides a *proof outline* of the formula

$\{I_1 \wedge I_2\}$ "LOOP BODY OF IBUFFER_FILL" $\{I_1 \wedge I_2 \wedge \text{input\_buffer} = \text{bus\_A}\}$

The proof is itself composed of proofs of more primitive formulas, for example, $\{I_1 \wedge I_2\}$ **await** ibuffer_full = 1 **do** ibuffer_full = 0 **od** $\{I_1 \wedge \text{ibuffer\_full} = 0 \wedge \text{ibuffer\_empty} = 0\}$. It uses the semantics of assignment, **await**, and **sig** statements, the rules of inference for statement sequences and parallel statements, and the rule of consequence.

The formula $I_1 \wedge I_2$ is the invariant for the **forever** loop body. It is initially true (i.e., when the loop is entered) as a result of the **init** statement, because of the initialized values of the synchronizer as shown in the declaration part of INPUT (Fig. 7.2). That it is an invariant is demonstrated by the above proof outline.

Consider now the procedure UNPACK.BUFFER_TRANSF. To prove

its correctness we need, in addition to I_2 defined earlier, the following assertions:

I_3: is_branch = 0
I_4: (obuffer_top_full = 0 $\wedge$ obuffer_top_empty = 1) $\vee$
  (obuffer_top_full = 1 $\wedge$ obuffer_top_empty = 0)

The proof outline for this procedure is shown as Fig. 8.2. The invariant for the **forever** loop is $I_2 \wedge I_3 \wedge I_4$. This clearly holds when the loop is first entered, as the initial values of the data objects indicate. The proof outline shows that the assertion holds after each iteration through the loop.

The invariant for the inner **repeat** loop is the assertion

$$\text{ibuffer\_full} = 0 \wedge \text{ibuffer\_empty} = 0 \wedge I_3 \wedge I_4 \wedge Q_1$$

Note that the semantics of the **repeat . . . until** statement (cf. Section 8.2.6, Appendix A) is such that this invariant holds after the first iteration through the loop. On termination of the **repeat** statement we have

$$\text{ibuffer\_full} = 0 \wedge \text{ibuffer\_empty} = 0 \wedge I_3 \wedge I_4 \wedge Q_1$$
$$\wedge (\text{ibuffer\_pointer} = 8 \vee \text{output\_buffer}[7].\text{lis} = 1)$$

which in turn implies the formula Q3:

$$(\text{ibuffer\_full} = 0 \wedge \text{ibuffer\_empty} = 0$$
$$\wedge I_3 \wedge I_4 \wedge \text{output\_buffer}[7].\text{full} = 1)$$
$$\wedge [(\text{ibuffer\_pointer} = 8 \wedge \text{output\_buffer}[7].\text{instrn} =$$
$$\text{input\_buffer}[\text{ibuffer\_pointer} - 1] \wedge$$
$$\text{output\_buffer}[7].\text{lis} = 0) \vee$$
$$(0 \leq \text{ibuffer\_pointer} < 8 \wedge \text{output\_buffer}[7].\text{instrn} =$$
$$\text{input\_buffer}[\text{ibuffer\_pointer}] \wedge$$
$$\text{output\_buffer}[7].\text{lis} = 1)]$$

In order to show that INPUT and UNPACK are interference-free we must prove the following. Let S_I and S_u denote the **forever** loop bodies in INPUT.IBUFFER-FILL and UNPACK.BUFFER_TRANSF respectively. Then

1. UNPACK does not interfere with INPUT if:
   (a) $\{\text{post}(S_I) \wedge \text{pre}(S_u)\} S_u \{\text{post}(S_i)\}$
   (b) For each non-**await** statement S'_I in S_I:
     $\{\text{pre}(S'_I) \wedge \text{pre}(S_u)\} S_u \{\text{pre}(S'_I)\}$

Proof of 1a.

Now

$$\text{post}(S_I): \{I_1 \wedge I_2 \wedge \text{input\_buffer} = \text{bus\_A}\}$$

We must show that each assignment in S_u leaves post(S_I) intact. The statement S_u has no effect on either "packet_sent" or "bus_A_empty." Hence I_1 is unaffected. The only statement involving "bus_A" or "input_buffer" uses the latter as a data source, hence the relation "input_buffer = bus_A" remains true. Finally, since I_2 is also in post(S_u), I_2 is true.

Proof of 1b.

We must show that each assignment in S_u leaves true each precondition in S_I. Now, the only data objects common to S_I and S_u are the synchronizers "ibuffer_full" and "ibuffer_empty," and the buffer variable "input_buffer." The only assignment in S_u that affects these is the last statement in S_u, namely, "**sig** ibuffer_empty." The only precondition in S_I that references "ibuffer_empty" is the clause I_2 in pre(S_I). Thus we must show that

$$\{I_2 \wedge \text{pre}(\textbf{sig ibuffer\_empty})\} \textbf{ sig ibuffer\_empty } \{I_2\}$$

This is clearly true, since I_2 is in post(**sig** ibuffer_empty).

2. INPUT does not interfere with UNPACK if:
   (a) $\{\text{post}(S_u) \wedge \text{pre}(S_I)\} S_I \{\text{post}(S_u)\}$
   (b) For each non-**await** statement S'_u in S_u

   $$\{\text{pre}(S'_u) \wedge \text{pre}(S_I)\} S_I \{\text{pre}(S'_u)\}$$

Proof of 2a.

post(S_u) = $\{I_2 \wedge I_3 \wedge I_4\}$. Of these clauses, I_3 and I_4 refer to data objects that are not referenced at all in INPUT (i.e., by S_I). Hence I_3 and I_4 will not be affected by the execution of S_I. The remaining clause I_2 is in post(S_I). Hence I_2 will also remain true.

Proof of 2b.

Again, we note that the only data objects common to S_I and S_u are the synchronizers "ibuffer_full" and "ibuffer_empty," and the variable "input_buffer." The only statements in S_I that affect these are "input_buffer := bus_A" and "**sig** ibuffer_full." The relevant preconditions in S_u are

proc BUFFER_TRANSF;
   forever do
      $\{I_2 \wedge I_3 \wedge I_4\}$

      await ibuffer_full = 1 do ibuffer_full := 0 od;
      {ibuffer_full = 0 $\wedge$ ibuffer_empty = 0 $\wedge$ I_3 $\wedge$ I_4}

      ibuffer_pointer := 0;
      {ibuffer_full = 0 $\wedge$ ibuffer_empty = 0 $\wedge$ ibuffer_pointer = 0 $\wedge$ I_3 $\wedge$ I_4}
      repeat
         {ibuffer_full = 0 $\wedge$ ibuffer_empty = 0 $\wedge$ I_3 $\wedge$ I_4} $\wedge$
   "Q1" {(0 $\leq$ ibuffer_pointer $\leq$ 8) $\wedge$ (output_buffer[7].full = 1) $\wedge$
         [(output_buffer[7].instrn = input_buffer[ibuffer_pointer] $\wedge$
            output_buffer[7].lis = 1) $\vee$
          (output_buffer[7].instrn = input_buffer[ibuffer_pointer - 1] $\wedge$
            output_buffer[7].lis = 0)]}

      await obuffer_top_empty = 1 do obuffer_top_empty := 0 od;
      {ibuffer_full = 0 $\wedge$ ibuffer_empty = 0 $\wedge$ obuffer_top_full = 0
         $\wedge$ obuffer_top_empty = 0 $\wedge$ I_3 $\wedge$ Q1}

      do output_buffer[7].instrn := input_buffer[ibuffer_pointer]
      □ output_buffer[7].full := 1
      od;

      {(ibuffer_full = 0 $\wedge$ ibuffer_empty = 0) $\wedge$ I_3 $\wedge$
       (obuffer_top_full = 0 $\wedge$ obuffer_top_empty = 0) $\wedge$
   "Q2" (0 $\leq$ buffer_pointer $\leq$ 8) $\wedge$ (output_buffer[7].full = 1)
         $\wedge$ (output_buffer[7].instrn = input_buffer[ibuffer_pointer])}
      sig store_re1_jump_store_start;
      await is_branch = 1 do is_branch = 0 od;
      output_buffer[7].lis := jump_trace_req_resp_line;

      {(ibuffer_full = 0 $\wedge$ ibuffer_empty = 0) $\wedge$ (obuffer_top_full = 0 $\wedge$
         obuffer_top_empty = 0) $\wedge$ I_3 $\wedge$ Q2}

         if output_buffer[7].lis = 0 => ibuffer_pointer :=
                                    ibuffer_pointer + 1 fi;

      {(ibuffer_full = 0 $\wedge$ ibuffer_empty = 0) $\wedge$ (obuffer_top_full = 0 $\wedge$
         obuffer_top_empty = 0) $\wedge$ Q1}

         sig obuffer_top_full;

Figure 8.2 Proof outline for UNPACK.BUFFER_TRANSF.

```
          {ibuffer_full = 0 ∧ ibuffer_empty = 0 ∧ I₃ ∧ I₄ ∧ Q1}
    until ibuffer_pointer = 8 ∨ output_buffer[7].lis = 1

"Q3" {ibuffer_full = 0 ∧ ibuffer_empty = 0 ∧ I₃ ∧ I₄ ∧
      [(ibuffer_pointer = 8 ∧ output_buffer[7].instrn =
        input_buffer[ibuffer_pointer - 1] ∧ output_buffer[7].full = 1 ∧
        output_buffer[7].lis = 0)  ∨
       (0 ≤ ibuffer_pointer < 8 ∧ output_buffer[7].instrn =
        input_buffer[ibuffer_pointer] ∧ output_buffer[7].full = 1 ∧
        output_buffer[7].lis = 1)]

      sig ibuffer_empty;

      {I₂ ∧ I₃ ∧ I₄}
    od
endproc
```

Figure 8.2 (*Continued*)

$$\text{pre}(S_u): I_2 \wedge I_3 \wedge I_4$$
$$\text{pre}(\textbf{sig ibuffer\_empty}): Q_3$$

Thus it must be shown that these preconditions are unaffected by the execution of the above-mentioned statements in S_1.

First, clearly, $\{I_2 \wedge I_3 \wedge I_4 \wedge \text{pre}(\text{input\_buffer} := \text{bus\_A})\}$ input_buffer := bus_A $\{I_2 \wedge I_3 \wedge I_4\}$ is true since input_buffer does not appear in $I_2 \wedge I_3 \wedge I_4$.

Second, the formula $\{I_2 \wedge I_3 \wedge I_4 \wedge \text{pre}(\textbf{sig ibuffer\_full})\}$ **sig ibuffer_full** $\{I_2 \wedge I_3 \wedge I_4\}$ is true, since post(**sig ibuffer_full**) $\supset I_2$, and I_3, I_4 remain unaffected by this **sig** statement.

Third, consider the formula

$$\{Q_3 \wedge \text{pre}(\text{input\_buffer} := \text{bus\_A}\} \text{ input\_buffer} := \text{bus\_A}\{Q_3\}$$

Note that Q_3 cannot be presumed to remain intact since "input_buffer" appears in this assertion. But since we are only concerned with the precondition of "**sig ibuffer_empty**" in S_u such that its postcondition holds, we can weaken the precondition so that

$$\text{pre}(\textbf{sig ibuffer\_empty}): Q'_3 = \text{ibuffer\_full} =$$
$$0 \wedge \text{ibuffer\_empty} = 0 \wedge I_3 \wedge I_4$$

yet the desired postcondition is still obtained, that is:

$$\{Q'_3\}\textbf{sig ibuffer\_empty}\{I_2 \wedge I_3 \wedge I_4\}$$

still holds. Clearly this is true. Hence, using this weakest precondition we see that

$$\{Q'_3 \wedge \text{pre}(\text{input\_buffer}:=\text{bus\_A})\}\text{input\_buffer}:=\text{bus\_A}\{Q_3\}$$

is true. This completes the proof showing that the two mechanisms INPUT and UNPACK are interference-free.

Consider now OBUFFER_EMPTY. When it enters the outer loop, its two interactions are with INPUT through synchronizers "obuffer_top_full" and "obuffer_top_empty," and with PROP through synchronizers "bus_B_full" and "bus_B_empty." This leads to the following assertions that will be part of the precondition of the **forever** loop body:

I_4: (obuffer_top_full = 1 $\wedge$ obuffer_top_empty = 0)
$\quad \vee$ (obuffer_top_full = 0 $\wedge$ obuffer_top_empty = 1)
I_5: (bus_B_empty = 1 $\wedge$ bus_B_full = 0)
$\quad \vee$ (bus_B_empty = 0 $\wedge$ bus_B_full = 1)

We also assume that bus_B contains some value B_0:

$$\text{bus\_B} = B_0$$

Let the initial contents of "output_buffer [0].instrn" be denoted by 0_0, that of output_buffer[1].instrn by $0_1, \ldots,$ the contents of output_buffer[7].instrn by 0_7. Then the precondition for the entire **forever** loop body (denoted as S_T) is given by

$$\text{pre}(S_T): I_4 \wedge I_5 \wedge \text{bus\_B} = B_0 \wedge I_6 \qquad (8.6)$$

where

I_6: output_buffer[0].instrn = 0_0 $\wedge \ldots \wedge$
$\quad$ output_buffer[7].instrn = 0_7

In each iteration through the outer loop body it must transfer from

1. output_buffer[0] to bus_B and signal that bus_B is full;
2. output_buffer[i] to output_buffer[i−1] for $6 \leq i \leq 1$;
3. output_buffer[7] to output_buffer[6] and indicate that output_buffer[7] is empty.

These transfers may be effected only when the following conditions are satisfied:

VERIFICATION OF DATAFLOW

1. Transfer 1 will take place only if output_buffer[0] is full and bus_B is empty. We do not want to transfer an "undefined" or "empty" instruction to bus_B.[1]
2. Transfer 3 must take place only if output_buffer[7] has been filled.

Since output_buffer[i−1] has to be emptied before output_buffer[i] is transferred to it, transfers 1, 2, and 3 must be performed in the sequence

$$\text{TRANSFER 1;} \qquad (8.7)$$
$$\text{TRANSFER 2;}$$
$$\text{TRANSFER 3;}$$

Consider TRANSFER 1. We want this to satisfy the following assertions. Given

$$\{\text{output\_buffer[0].full} = 1 \wedge \text{bus\_B} = B_0\}$$

then

$$R1: \{\text{bus\_B} = 0_0 \wedge \text{bus\_B\_empty} = 0 \wedge \text{bus\_B\_full} = 1\}$$

Given

$$\{\text{output\_buffer[0].full} = 0 \wedge \text{bus\_B} = B_0\}$$

then

$$R2: \{\text{bus\_B} = B_0\}$$

This leads to the following proof outline for the first part of the **forever** loop body S_T (Fig. 7.4):

$\{\text{pre}(S_T)\}$
if output_buffer[0].full = 1 => {output_buffer[0].full = 1}

$$\textbf{await } \text{bus\_B\_empty} = 1 \textbf{ do}$$
$$\text{bus\_B\_empty} := 0 \textbf{ od};$$

$$\{\text{bus\_B\_empty} = 0 \wedge \text{bus\_B\_full} = 0\}$$

---

[1]This will only happen when the system is started up and until the first instruction from input_buffer has been passed to output_buffer[0].

$$\text{bus\_B} := \text{output\_buffer}[0].\text{instrn};$$

$$\textbf{sig } \text{bus\_B\_full};$$

fi $\{R_1\}$

$\{(R_1 \lor R_2) \land I_5\}$

Consider next TRANSFER 2. This may be represented by the following proof outline:

$\{(R_1 \lor R_2) \land I_5 \land I_6\}$
obuffer_ptr := 1 □ output_buffer [0].full := 0
$\{(R_1 \lor R_2) \land I_6 \land \text{obuffer\_ptr} = 1 \land \text{output\_buffer}[0].\text{full} = 0\}$

repeat
    $\{(R_1 \lor R_2) \land I_7\}$
    output_buffer[obuffer_ptr − 1] :=
        output_buffer[obuffer_ptr] □
        output_buffer[obuffer_ptr].full := 0;
    obuffer_ptr := obuffer_ptr + 1
until obuffer_ptr = 7

$\{(R_1 \lor R_2) \land I_5 \land I_7 \land \text{obuffer\_ptr} = 7\}$

where the invariant relation for the **repeat** statement is:

$I_7 : 1 \leq \text{obuffer\_ptr} \leq 7 \land$
  output_buffer[0].instrn = $0_1 \land \ldots \land$
    output_buffer[obuffer_ptr − 2].instrn = $0 \land$
      output_buffer[obuffer_ptr − 1].instrn = $0_{\text{obuffer\_ptr}-1} \land$
        $\ldots \land$ output_buffer[7].instrn = $0_7 \land$
        output_buffer[obuffer_ptr − 1].full = 0

The assertion I_7 is obviously true when the **repeat** statement is entered, since initially we have the condition

$$\text{obuffer\_ptr} = 1 \land \text{output\_buffer}[0].\text{full} = 0 \land I_6$$

which implies I_7.

Let the assertion be true after the $(i - 1)$th pass through the loop body ($1 \leq i \leq 7$). In that case

$$\text{obuffer\_ptr} = i$$

and the proof outline for the loop body itself is:

$\{$obuffer_ptr $= i(1 < i \leq 7) \wedge$
  output_buffer[0].instrn $= 0_1 \wedge \ldots \wedge$
    output_buffer[i $-$ 2].instrn $= 0_{i-1} \wedge$
      output_buffer[i $-$ 1].instrn $= 0_{i-1} \wedge \ldots \wedge$
        output_buffer[7] $= 0_7 \wedge$
  output_buffer[i $-$ 1].full $= 0\}$

output_buffer[obuffer_ptr $-$ 1] := output_buffer[obuffer_ptr] □
output_buffer[obuffer_ptr].full := 0;

$\{$obuffer_ptr $= i(1 \leq i \leq 7) \wedge$
  output_buffer[0].instrn $= 0_1 \wedge \ldots \wedge$
    output_buffer[i $-$ 2].instrn $= 0_{i-1} \wedge$
      output_buffer[i $-$ 1].instrn $= 0_i \wedge$
        output_buffer[i].instrn $= 0_i \wedge \ldots \wedge$
          output_buffer[7] $= 0_7 \wedge$
  output_buffer[i].full $= 0\}$

$\qquad$ obuffer_ptr := obuffer_ptr $+$ 1

$\{$obuffer_ptr $= i + 1(1 \leq i \leq 7) \wedge$
  output_buffer[0].instrn $= 0_1 \wedge \ldots \wedge$
    output_buffer[obuffer_ptr $-$ 2].instrn $= 0_{\text{obuffer\_ptr}-1} \wedge$
      output_buffer[obuffer_ptr $-$ 1].instrn $= 0_{\text{obuffer\_ptr}-1} \wedge$
        $\ldots \wedge$ output_buffer[7].instrn $= 0_7 \wedge$
  output_buffer[obuffer_ptr $-$ 1].full $= 0\}$

which is nothing but the invariant I_7.
  Finally, when the loop terminates

$\qquad I_7 \wedge$ obuffer_ptr $= 7 \supset$
  output_buffer[0].instrn $= 0_1 \wedge \ldots \wedge$
    output_buffer[5].instrn $= 0_6 \wedge$
      output_buffer[6].instrn $= 0_6 \wedge$
        output_buffer[7].instrn $= 0_7$

That is, TRANSFER 2 (see [8.7]) has been completed.
  This leaves TRANSFER 3. The proof outline for this segment is

$\qquad \{I_4 \wedge I_5 \wedge (R_1 \vee R_2) \wedge I_7 \wedge$ obuffer_ptr $= 7\}$

await obuffer_top_full $= 1$ **do** obuffer_top_full $:= 0$ **od**;
output_buffer[6] := output_buffer[7];
sig obuffer_top_empty

```
proc OBUFFER_EMPTY;
   forever do
      {I_4 ∧ I_5 ∧ I_6 ∧ bus_B = B_0}
      if output_buffer[0].full => {output_buffer[0].full = 1}
                                  await bus_B_empty = 1 do bus_B_empty := 0 od
                                  {bus_B_empty = 0 ∧ bus_B_full = 0}
                                  bus_B := output_buffer[0].instrn;
                                  sig bus_B_full
                                             {R1}
      fi
      {(R1 ∨ R2) ∧ I_5 ∧ I_6}
      do obuffer_ptr := 1 □ output_buffer[0].full := 0 od;
      {(R1 ∨ R2) ∧ I_6 ∧ I_5 ∧ obuffer_ptr = 1 ∧ output_buffer[0].full = 0}
      repeat
         {(R1 ∨ R2) ∧ I_5 ∧ I_7)}
         do output_buffer[obuffer_ptr-1] := output_buffer[obuffer_ptr]
            □ output_buffer[obuffer_ptr].full := 0
         od;
         obuffer_ptr := obuffer_ptr + 1
      until obuffer_ptr = 7 ;
      {(R1 ∨ R2) ∧ I_5 ∧ I_7 ∧ obuffer_ptr = 7}

      await obuffer_top_full = 1 do obuffer_top_full := 0 od;
      output_buffer[6]. := output_buffer[7];
      sig obuffer_top_empty;

         {I_4 ∧ I_5 ∧ (R1 ∨ R2) ∧ R3}
   od
endproc
```

Figure 8.3 Proof outline for OUTPUT.OBUFFER_EMPTY.

$$\{I_4 \wedge I_5 \wedge (R_1 \vee R_2) \wedge R_3\}$$

with

R_3 : output_buffer[0].instrn = $0_1 \wedge \ldots \wedge$
      output_buffer[6].instrn = $0_7 \wedge$
      output_buffer[7].instrn = 0_7

The proof for the complete procedure OBUFFER_EMPTY is outlined in Fig. 8.3.

What remains is to show that UNPACK.BUFFER_TRANSF and OUT-

PUT.OBUFFER_EMPTY are interference-free. We leave this as an exercise for the reader.

Remarks. In the foregoing discussion we have dealt strictly with the problem of proving partial correctness of the individual procedures. Since all three mechanisms, once activated, remain active forever, the question of termination of the mechanisms does not arise. However, both UNPACK.BUFFER_TRANSF and OUTPUT.OBUFFER_EMPTY contain within the **forever** loops repetition statements whose terminations need to be ensured to guarantee total correctness of the respective **forever** loop bodies S_u and S_T.

It is easy to see that the two **repeat** statements, one each in S_u and S_T, will both terminate. In the first case, a precondition of the **repeat** loop is "ibuffer_pointer = 0." Within the loop body, ibuffer_pointer is incremented by one each time the condition "output_buffer[7].lis = 0" is met. As long as this condition is true, ibuffer_pointer will continue to be incremented by one until eventually the terminating conditions "ibuffer_pointer = 8" becomes true; if, in any pass through the **repeat** body "output_buffer[7].lis ≠ 0," though ibuffer_pointer is not incremented by one, the terminating condition "output_buffer[7].lis = 1" will be met. In either case the **repeat** statement will terminate.

In the case of the inner loop in S_T, the **forever** loop body in OUTPUT.OBUFFER_EMPTY, the **repeat** is entered with the condition "obuffer_ptr = 1." In each pass through the **repeat** loop body, obuffer_ptr is incremented by one. Eventually the terminating condition "obuffer_ptr = 7" will be met and the loop will terminate.

8.5 BIBLIOGRAPHIC REMARKS

S. Owicki and D. Gries first described their proof technique in Owicki and Gries (1976). Gries (1976b) also applied this technique to the proof of an "on-the-fly" garbage collector. For other discussions of parallel programs and proofs of their correctness see Hoare (1976), Gries (1981), and Lamport (1979).

NINE

TOWARD HIGH-LEVEL MICROPROGRAMMING

In Chapter 5 I had suggested that one possible approach to the formal multilevel design of computer architectures is the use of a family of closely related design languages, each tailored to a specific range of architectural levels. The S* family (Dasgupta and Olafsson [1982]) is intended to constitute one such group of languages.

Along with S*A, the other complete member of this family (at this time of writing) is the microprogramming language schema S* (Dasgupta [1980b]). Given a machine M specified at the microarchitectural level, S* may be instantiated to a particular high-level, machine-specific microprogramming language S*(M), which may, in turn, be used for implementation of target exoarchitectures or (less commonly) endoarchitectures on M. If the target architecture is itself specified in S*A, the transformation from one description to the next may, from an intellectual point of view, be expected to be more tractable than if two totally unrelated languages were used. We say that S*A and S* (and, by implication, S*A and the subfamily of languages instantiated from S*) are *kin* languages.

One may ask why a language schema rather than a particular language has been designed for the purpose of high-level microprogramming. This question is addressed in this chapter, while the structure of S* is described in Chapter 10.

9.1 PROBLEMS OF HIGH-LEVEL MICROPROGRAMMING

It might be expected that the principles of programming language design would, with minor modifications, be applicable to the development of microprogramming languages. However, the latter poses some critical problems that are not ordinarily encountered in the software domain. Some of the more important of these problems are:

1. The presence of low-level parallelism.
2. The general belief that microprograms must be optimal, and that code generated from high-level language sources will compare poorly with hand-optimized code.
3. The variability in microarchitectures.
4. The fact that a microprogram, by definition, creates some desired target architecture on a specific microarchitecture.

9.1.1 Low-Level Parallelism

As a study of some of the earlier microprogrammed systems will indicate, microprograms form one of the earliest instances of parallel programs.

The nature of microparallelism is in general twofold. First there is the parallelism between microoperations executing from the same microinstruction. A microprogram in a state of execution can then be regarded as a single quasisequential process in which each member of the sequence is a collection of microoperations executing concurrently, usually within a microcycle (Fig. 9.1).

This kind of microparallelism (I will simply designate it Type I) is quite common, since it is present even in single processor systems. A recent and interesting example of a machine that exploits this type of low-level parallelism to create a single-instruction-stream/multiple-data stream (SIMD) system is the QA-1, designed at Kyoto University by Hagiwara et al. (1980).

The second kind of parallelism is that between communicating microprocesses, each being executed by physically distinct processors (Fig. 9.2). This type of microparallelism (I shall call it Type II) is possible, for example, on a network of microprogrammable processors, each running its own microprogram, but possibly sharing certain resources and therefore needing to communicate and exchange signals with other processors.

Most proposals for high-level microprogramming languages (with the possible exception of S*) have simply ignored Type II parallelism. However, because of its similarity to cooperating and communicating software systems, one can, as we shall see later, use concepts from parallel programming and adapt them to the microprogramming context.

Type I parallelism has proved to be surprisingly intractable from the point of view of representation. One solution to this problem is to simply ignore

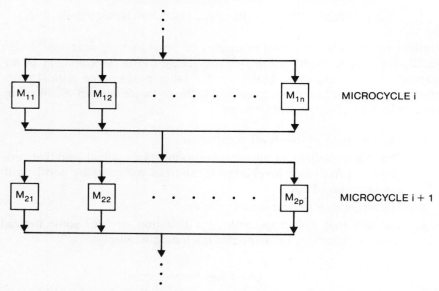

Figure 9.1 Quasisequential microprogram (copyright © 1980, ACM; reprinted with permission).

it—a microprogram would then be developed in a strictly sequential form, leaving to the compiler the task of automatically detecting parallelism, compacting the code, and producing a sequence of microinstructions that is functionally equivalent to the original source program.

This approach has prompted extensive study of the *microcode compaction problem*—the problem of automatic detection of parallelism in a sequential microprogram and consequent generation of microinstructions. However, despite the reasonable success that has been attained in automatic parallelism detection and code compaction, given the fundamental nature of microparallelism, it seems that the microprogrammer should have the choice of exercising control over this parallelism. This seems particularly so when one realizes that even the most efficient of the existing algorithms produce suboptimal microcode. Thus, although automatic microcode compaction provides one approach to coping with microparallelism, the design of language constructs for representing Type I parallelism explicitly seems necessary.

9.1.2 Optimality of Microcode

Since the quality of the microcode determines the performance of the computer at the exoarchitectural level, constructing the most efficient microprogram possible has traditionally been the overriding objective of microprogrammers. A major impediment to high-level microprogramming has

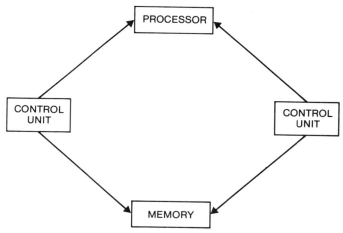

Figure 9.2 Communicating microprocesses.

been the generally held belief that code generated from a high-level microprogramming language would perform poorly compared to hand-produced and hand-optimized code.

It should be noted, at this point, that the term "optimization" is used in various ways in the microprogramming literature. Generally speaking, a microprogram may be said to be optimal if there are no other functionally equivalent microprograms that can run on the same machine and require a smaller number of clock cycles.

A major portion of the work on microprogram optimization to date has been concerned with compaction, a problem mainly associated with horizontal microprogramming systems.[1] In contrast, vertical optimization is concerned with various kinds of code transformations, such as the elimination of redundant code or useless statements (and is therefore, similar in its objectives and techniques to program optimization strategies).

The earliest studies of vertical optimization were performed by Kleir and Ramamoorthy (1971). However, very little work has been reported in the open literature in the development of optimizers for "real" machines, one of

---

[1] The term *horizontal* in this context is used to denote microinstruction word organizations that exhibit one or more of the following characteristics: (1) they enable different functional units to be controlled independently in the same microinstruction; (2) they lead to relatively large microinstruction word lengths (of the order of 64–360 bits); (3) they allow the exercise of control at the level of microoperations.

The term *vertical* is used in the literature in at least two ways: (1) it is often used to refer to microinstructions that encode for single microoperations; (2) it also denotes microinstruction word organizations that lead to small word lengths, of the order of 16–32 bits. In general, there is very little parallelism possible within the individual microinstructions. Thus horizontal and vertical schemes provide a useful if somewhat imprecise taxonomy for control store organizations.

the few exceptions being Patterson's (1978, 1981) experiments with the STRUM system.

These experiments were concerned with the comparative efficiencies of hand-produced and mechanically generated microcode. In one particular study, two emulations of the Hewlett Packard HP2115 architecture were constructed for the Burroughs D Machine. The first version was implemented in a conventional manner—that is, writing the emulator in TRANSLANG (a microassembly language for the D Machine) and testing and debugging the microcode directly on the D Machine. The second version was written in a Pascal-like microprogramming language called STRUM and was compiled, optimized, and verified by the STRUM compiler. The STRUM emulator was observed to be both faster and slightly smaller. Patterson conjectured that this unexpected result was a consequence of the relatively large size of the emulator: while for smaller examples (10–20 microinstructions) it is possible for a programmer to spend considerable periods of time hand-optimizing the code, the use of an automated, global optimizer provides distinct advantages in the case of larger microprograms, since their size may render them intellectually unmanageable otherwise.

While the results of all such experiments involving the measurement of human performance has to be interpreted with a great deal of caution, it is fair to say, I think, that such studies at least indicate the real practicability of automated optimizers that are capable of producing code roughly comparable in efficiency to hand-produced code.

The problem of microcode compaction has proved to be far richer in its scope and yield than that of vertical optimization. In essence, most of the workers in this field regard the problem of local compaction (i.e., compaction of straight-line microprogram segments) to be a solved problem, while the theory of *global* compaction (encompassing the analysis of microprograms that include iterations and conditionals) is far less developed, though some work has been reported on this more general problem. Again, the actual application of the theory of microcode compaction to the design of practical compactors has been sparse—with a few notable exceptions.

Rideout's (1981a, 1981b) implementation of the so-called "linear" local code compaction algorithm (Dasgupta and Tartar [1976], Landskov et al. [1980]) on the QM-1 and the related work by Ma and Lewis (1980, 1981) for the PDP-11/40E provide important evidence that merely applying local compaction techniques may in itself produce object microcode to within 25% of the optimal. With further compaction through global techniques augmented with vertical optimization strategies, code that is within 10% of the optimal may result.

To assess the efficacy of global compaction, Grishman and Bogong (1982) conducted studies on microcode produced for a horizontally organized machine called PUMA (Grishman [1980]). In their experiments, carefully hand-coded PUMA microcode served as a benchmark. The code had been reviewed by several readers and was consequently thought to be optimally compacted.

PROBLEMS OF HIGH-LEVEL MICROPROGRAMMING

The global compaction algorithm tested was Fisher's "trace scheduler" (Fisher [1981], Fisher et al. [1981]), which was implemented in SETL running on a VAX 11/780. To provide input to the compactor, the hand-coded sequences were rewritten in sequential form. Finally, to compare Fisher's global compactor with local compaction, manual local compaction of the code was also performed.

For the code segments under test, the output of the global compactor compared quite favorably with hand-compacted code. For example, the average execution time (in cycles) for a floating point divide routine required 16.1 cycles and 16.4 cycles respectively when hand and globally compacted. The initialization part of a floating point multiply required 14 cycles in both cases. A hand-compacted floating point add performed slightly better than its globally compacted counterpart. Finally, in these particular set of experiments, local compaction performed rather poorly, requiring some 60% additional cycles.

Thus, as in the case of Patterson's studies, these results provide further evidence that automated compaction is indeed feasible, and is capable of producing output that is competitive with hand-compacted code.

9.1.3 Machine-Specificity of Microprograms

The problem of machine specificity is rooted in a fundamental distinction between programs and microprograms. A program, written in any given programming language, may be made to run on an arbitrary computer by interpreting, if necessary, a *virtual* interface between the program and the computer (Fig. 9.3). A microprogram, on the other hand, can only be interpreted by a *real* interface—that is, by an interface (architecture) defined directly by physical hardware (Fig. 9.4).

In the context of high-level microprogramming, this explains, in part, the difficulty of transferring the concept of a high-level (machine-independent) programming language directly into the microprogramming domain. We can always guarantee that a programming language L_p can be implemented by the simple (though possibly inefficient) expedient of defining an appropriate virtual machine that executes L_p programs. This virtual machine itself is implemented at a lower level on a real machine. This is, for example, the basis for the implementation by Brinch Hansen (1977) and Hartmann (1977) of the Concurrent Pascal language on the PDP 11/45—a virtual Concurrent Pascal machine executes (compiled) Concurrent Pascal programs, and the virtual machine is itself interpreted in PDP 11 machine language code.

To talk of a microprogram or an emulator running on a virtual machine in this same fashion is, however, operationally meaningless. One must be able to ensure that a microprogramming language L_m can be implemented on a given microarchitecture without recourse to some intermediate abstract architectural level.

It is for this reason that, in this writer's view, the issue of emulator or microprogram *portability*—a design objective, as we shall see below, for

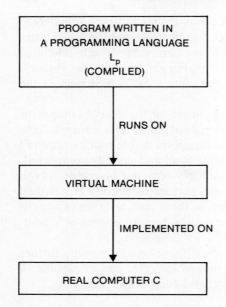

Figure 9.3 Program executing on a virtual machine.

several language proposals—is somewhat of a red herring, in the following sense. Let E be an emulator that is transportable from one host machine to another. Then, of necessity, E must be a description of a target (exo)architecture, independent of the nature of the host machine. It is then no longer a microprogram (which expresses a relationship between the target and a real host), but simply a description of the target architecture. It is then sufficient to describe such architectures in an architectural, rather than a microprogramming, language.

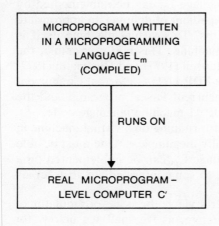

Figure 9.4 Microprogram executing on a real machine.

PROBLEMS OF HIGH-LEVEL MICROPROGRAMMING 117

> **State components:** control memory, registers, buses, local store, stack, flags
> **State transition mechanisms:** microoperations
> **Organizing mechanisms:** microinstructions, sequencing logic, address generation mechanisms, timing, subroutine mechanisms

Figure 9.5 Microarchitectural components (copyright © 1980, ACM; reprinted with permission).

It must be pointed out, however, that even if we were to dismiss portability, a truly machine-independent language—that is, one in which one may write microcode for different host microarchitectures—would be most desirable in the context of VLSI-based processors. For example, Li (1982) has made the observation that the microarchitecture may be modified frequently during the VLSI system development process, with attendant changes to the microcode. Hence, if a high-level language were to be used in this design process, it seems almost imperative for it to be independent of any particular microarchitecture. Note that the resulting high-level microprograms must still be machine-specific. At this writing, however, it is not clear whether such languages are feasible.

9.1.4 Variability of Microarchitectures

This constraint follows quite naturally from the problem of machine-specificity. Indeed, the problems induced by the variability of microarchitectures result from the fact that we cannot express microprograms without reference to some underlying host machine. Thus, a machine-independent microprogramming language will have to be sufficiently general to cope with the variations in microarchitectures.

Generally speaking, we can characterize the microarchitecture in terms of the components shown in Fig. 9.5. Of the three broad categories indicated, the state component includes entities that collectively store the state of the microprogram-level machine. State transitions are effected basically by microoperations, which are the most primitive actions available at this level. The precise ways in which microprograms can be constructed are dictated by the set of organizing mechanisms defined for the interface.

Clearly, all three categories are sources of micro-architectural variability. But the most critical of these appears to relate to the semantics of microoperations and the timing of microinstructions.

Example 9.1

The Varian 75 (Varian [1975]), a horizontal user-microprogrammable system, includes among its functions an operation A-B-1 performed by its ALU, where A and B are the two input buses to the ALU. The result of this operation is placed on the ALU output bus. There is, however, a fourth input to the ALU that is not shown explicitly: the ALU carry-in bit, which can be set under program control to 0, 1, the value of the ALU carry flag, or its complement. Thus, the semantics of this particular microoperation are partly determined by the state of a hidden variable.

This is not an unusual situation, since in many other systems ALU operations have the value of the carry-in bit as an implicit input. However, this is not always true; for example, in the Microdata 1600 (Microdata [1970])—a vertically organized microprogrammable computer—the input from the carry bit has to be explicitly specified.

Example 9.2

The Varian 75 performs a large number of arithmetic and logic functions. These operations result in the following condition flags being set: carry, sign, all-ones, overflow, all-zeros. In the case of the Cal Data Processor (Agrawala and Rauscher [1976]), an ALU operation generates the condition codes carry out, overflow, zero result, negative, positive, and odd result. In the Microdata 3200 an ALU operation records only the following conditions: zero result, negative carry, and overflow (Agrawala and Rauscher [1976]). Quite clearly, the precise semantics of the corresponding sets of microoperations differ.

Example 9.3

The Nanodata QM-1 (about which I will have more to say in Chapter 11) has a two-level microprogram store, of which the lower level is termed the nanostore and the corresponding program the nanoprogram (Nanodata [1979]). A highly idiosyncratic nano-operation (i.e., a micro-operation at the nanoprogram level) is WRITE NS, which is available for writing data into nanostore. If an illegal byte of the addressed word in nanostore (>19) is specified by the programmer, the write operation does not take place; instead the addressed word is *read* from nanostore.

A particularly problematic area is the timing characteristics of microoperations and instructions, since these vary widely between machines. The taxonomy of microarchitectures conveniently categorizes systems as being either *monophase* or *polyphase*, where the former term refers to the control of all micro-operations executing within a microinstruction by a single-phase clock cycle and the latter to control by a multiple-phase clock cycle. However, polyphase systems may vary widely in their timing characteristics, as may monophase ones.

Example 9.4

In the Varian 75 the memory fetch operation

$$mir := mem[pc]$$

requires more than one microcycle. Here pc, the program counter, is the memory address source. However, if this operation is initiated in microcycle i, the operation "pc := pc + 1" can be performed in microcycle $i + 1$ (even though the memory operation may not have been completed).

Example 9.5

An unusual timing feature is also present in the QM-1. In this system the 360-bit nanoinstruction is composed of five 72-bit vectors K, T_1, T_2, T_3, and T_4, such that at any one time, state transition at the nanoprogrammable machine level is controlled by the K and one of the T vectors. Each of the T vectors normally remains active for one "T period": 80 nanoseconds. However, by the use of a special T-vector field, this period can be doubled to allow certain operations to complete before the activation of the next T vector.

9.2 SOME APPROACHES TO HIGH-LEVEL MICROPROGRAMMING

9.2.1 Machine-Specific Languages

Clearly, the most straightforward approach to resolving the problem is to develop and use machine-oriented high-level microprogramming languages. An example is STRUM, an Algol-like language designed and implemented by Patterson (1976) that is oriented toward the Burroughs D Machine. A particularly notable contribution of the STRUM project was the introduction of *microcode verification* issues as a factor in the design of the language. Patterson's main objective was to construct a language and a programming system that would permit well-structured, verifiable microprograms to be written for the Burroughs D Machine without having to sacrifice object code efficiency. Thus Patterson paid a great deal of attention to the structuring capabilities in the language, which has, as a result, four distinct repetition statements, four selection statements, a macrostatement, and the procedure construct. The data structuring capabilities are, however, quite simple.

9.2.2 Abstract Machine Languages

Given the machine-specific nature of microprograms and the verifiability of microarchitectures, one possible machine-independent solution to the problem is a language defined for some abstract microarchitecture. A language named VMPL, proposed by Malik and Lewis (1978), is based on this approach. A program written in this language is essentially an abstract microprogram within which the declared variables refer to objects in the target (emulated) machine, and all specified executable statements result in transformations in the states of these declared variables. A program in VMPL is thus a specification of some machine exoarchitecture, in much the same way that S*A or ISPS are used for describing architectures. In order to execute this microprogram on some given host machine, the program is first translated into an intermediate language and then compiled into microcode. Clearly, compiling source microprograms into efficient object code will involve the use of complex scheduling, allocation, and optimization algorithms.

In a recent review of the status of high level microprogramming, Davidson (1983) has observed that the specific context in which microcode is being written could largely determine whether or not a compiler is economically justifiable. For example, given a user-microprogrammable computer on which many different instruction sets are to be emulated, the development of a compiler may seem necessary. However, in designing a single chip processor in which microcoding overlaps other phases of the project, the latter may be unduly delayed because of the compiler development effort. Further, as noted in the preceding section, the microarchitecture may itself evolve during the project, thereby necessitating frequent changes to the code generation and optimization phases of the compiler.

Davidson makes the interesting point that in such contexts, a more appropriate, technically feasible, and economically sound approach is to associate a design *methodology* with the high-level language. Indeed, it has been reported by both Davidson (1982) and Li (1982) that microcode for the Intel iAPX 432 (Intel [1981]) was developed along these lines. The high-level algorithms were first written in ADA, without any binding of data objects or operations to specific hardware. The ADA description was then hand-translated into a lower-level representation by using a subset of ADA, with some attendant hardware bindings. Finally, further (manual) translation into assembly-level microcode was performed.

Clearly, the use of a family of related languages as envisioned in Chapter 5 is relevant to this approach—with the added qualification that the design process in this case would extend down to the object (or assembly) code level. From a methodological viewpoint, the interesting problem here is to demonstrate the correctness of the transformations between design stages.

9.2.3 Extensibility

An alternative way of achieving machine independence is to use the concept of language extensibility; this formed the basis for DeWitt's (1976a, b) language, EMPL.

The EMPL language is defined basically in terms of a *core language* and a set of facilities that allow the programmer to extend the core. The facilities are of two types—*extension statements* and *extension operators*. The former, like the class concept in SIMULA (Dahl and Nygaard [1970]) or the monitor concept in Concurrent Pascal (Brinch Hansen [1977]), allow the user to define new data types and operations to manipulate instances of the data types. Extension operators are used simply to define new operators in the (extended) language. Thus, extension statements are useful in order to define objects and associated microoperations that are present in the host machine but not in the core language. Extension operators serve to define new microoperations.

Using this approach the programmer can customize the language to fit a particular microarchitecture. Thus, if the architecture supports stack-manipulating microoperations, EMPL can be extended by declaring a new type called STACK together with the operations defined on it.

The EMPL language thus provides the microprogrammer with a capability for developing machine-specific microprograms using a machine-independent language. It also provides a mechanism for transporting microprograms from one machine to another in the following way.

Assume that a program has been written for a host machine M that supports a hardware stack. The programmer would then provide a STACK extension statement in his or her program. When this is compiled to run on M, each occurrence of the object code equivalent of a stack operation (e.g., PUSH) would be compiled into a single microoperation.

Suppose, however, the same program were to be run on machine M', which does not have a hardware stack. A control statement preceding the source program would indicate this fact. This would cause the stack storage to be simulated (in main store) and a stack operation would be compiled into a sequence of microoperations corresponding to the original EMPL code.

The EMPL language has, to my knowledge, never been implemented. A more recent language called MARBLE, designed by Davidson and reported in Davidson and Shriver (1980, 1981) and Davidson (1980), had rather similar objectives, although the means of achieving them were different. Among other things, MARBLE is strongly influenced in its syntax by Pascal, and possesses richer data structuring capabilities than EMPL.

As with EMPL, the primary goal in the design of MARBLE was to provide a medium in which microprograms could be written either from some abstract machine or for specific host architectures. In the former case the problems discussed earlier of automatic resource binding and code optimization clearly have to be resolved. Of immediate interest here, however, is the means by which machine-specific information is provided at the source program level.

The MARBLE language allows the representation of host machine data objects in terms of standard type constructors. Similarly, MARBLE functions can be used to represent machine-specific functional resources. These constructs can be recognized by the compiler and passed on to the code generator.

Basically, the MARBLE compiler first partitions the source program into a machine-independent part and a machine-dependent part. The former is translated into a machine-independent intermediate form which is then passed to the code generator, where its resources are bound and merged with the machine-dependent code.

As mentioned above, MARBLE is based on Pascal. However, some additions were made to facilitate resource binding at the source level. As a specific example, consider the definition of an optional multiply unit. Assume that its inputs are contained in two 16-bit registers, mplr and mplnd, and that the product is to be placed in a 32-bit register called "result."

If this unit is actually present in the processor the following code may be written:

type %mplr, %mplnd = **bit** 16;
    %result = **bit** 32;

procedure bound multiply (**in** a: %mplr; **in** b: %mplnd;
                      out c: %result);
    begin
        "multiply code"
    end

The registers are declared as *bound types* (denoted by the percent sign). The procedure "multiply" is similarly bound to the hardware multiply unit. Finally, the procedure parameters are bound to the registers mplr, mplnd, and result. Given such a declaration, when a call to this procedure is encountered, the compiler ignores the code in the procedure body and simply translates the call into the appropriate microoperation.

If a hardware multiply unit is not present in the processor, this procedure must be unbound. This can be done by inserting suitable compiler directives before and after the procedure declaration. Subsequently, calls to the procedure compiles into a call to the microsubroutine that implements the procedure body.

An implementation of MARBLE for the Microdata 3200 [Agrawala and Rauscher (1976)] exists. At this writing, further implementation studies for a more horizontal machine are being carried out at the University of Southwestern Louisiana.

9.2.4 Machine-Independent Schemata

I shall now describe an approach that is developed in more detail in the next two chapters. This approach is based on the concept of a microprogramming language schema. While I tend to regard the schema approach as complementing DeWitt's (1976b) extensibility approach, I feel that the use of schemata and their instantiations (as explained below) provides a simpler solution to the microprogramming language problem than is obtained by the use of extensible languages.

Basically, the issues underlying the schema are those discussed in the previous section. But what really emerges from this discussion, and from an evaluation of some of the preceding language proposals, is that given the nature and intent of microprogramming, the criteria by which we may designate a microprogramming language as high-level are quite different from the criteria we use for programming languages. That is, we may be content to designate a microprogramming language as high-level if it possesses some particular machine- and implementation-independent characteristics, though the language may appear quite low-level and primitive compared to a typical programming language. We may identify these essential characteristics as the following:

1. The ability to construct control structures for designating clearly, and without ambiguity, both sequential and parallel flow of control.
2. The ability to describe and name, arbitrarily, microprogrammable data objects or parts of such data objects.
3. The ability to construct microprograms whose structure and correctness can be determined and understood without reference to any control store organization.

A language that possesses these characteristics may not be directly applicable to all microarchitectures. But it may provide a fairly detailed language skeleton containing constructs that are common to a very large range of machines.

This leads to the concept of a microprogramming language schema. A schema S in fact denotes a family of languages, such that for a given microarchitecture M, a particular language S(M) is obtained by extending or filling in the (partial) specifications of S by using the specific, idiosyncratic properties of M. In other words, S is instantiated into a particular language S(M) on the basis of (or with respect to) M.

Given the nature of microprogramming, the schema approach offers, if not an optimal solution to the microprogramming language problem, a solution that is "satisficing"—to use Simon's (1981) term.

Apart from using extensible languages, schemata appear to represent the best we can achieve in terms of machine-independent languages. It is expected that a good schema S should also possess the following properties.

1. Since all the possible construct and data types are defined in S, given a reasonable microarchitecture M, instantiating S to S(M) should not require any new construct or data types.
2. Given two languages $S(M_1)$ and $S(M_2)$ instantiated from S, the only distinction (if any) should be at the level of the elementary statements.

The precise relationship between a schema, its instantiation, and the actual microprogramming interface is best described by considering their respective levels of abstraction. Figure 9.6 shows the microprogramming domain at three levels of abstraction. Level 0 is the same as the interface

> **Level 2**
>
> State components—microprogram data types (e.g. **bit, sequence, array**, etc.).
> State transition mechanisms—primitive action types (e.g. assignments).
> Organizing mechanisms—concurrent action, action-within-a-microcycle, sequential action, iteration, selection, procedures.
>
> **Level 1**
>
> State components—microprogrammable data objects (variables and constants).
> State transition mechanisms—primitive actions defined on specific data objects.
> Organizing mechanisms—concurrent action, action-within-a-microcycle, sequential action, iteration, selection, procedures.
>
> **Level 0**
>
> State components—control memory, registers, buses, local store, flags.
> State transition mechanisms—microoperations.
> Organizing mechanisms—microinstructions, sequencing logic, address generation mechanisms, timing, subroutine mechanisms.

Figure 9.6 Abstraction levels for the microprogramming domain (copyright © 1980, ACM: reprinted with permission).

shown in Figure 9.5; as mentioned before, conventional microprograms are designed for this level. Notice that this requires detailed knowledge on the part of the microprogrammer of the control store organization. This includes the encoding scheme (format) for microinstructions, the logic for sequencing microinstructions, and timing characteristics.

A major simplification results when we abstract from this level to one where the control store organization is partially transparent (level 1). The microprogrammer must still have knowledge of the (machine-specific) state transition mechanisms. However, their hardware characteristics may be suppressed so that at level 1 we obtain as state components (machine-specific) variables and constants, and as transition mechanisms, primitive actions defined on these data objects. The main organizing mechanisms at this level are various forms of control structures.

If we abstract still further we obtain classes of data objects and actions, with each such class denoting a particular data or action type. The microprogramming domain is characterized at this level (level 2) in essentially machine-independent terms. The microprogramming language schema S* described in the following chapter is defined at level 2. Any particular instantiation will result in a language associated with a level 1 machine. Finally, the task of a compiler is to transform programs written in the instantiated language to programs expressed in a level 0 language.

It is worthwhile comparing, briefly, the schema and MARBLE approaches. Both are concerned with the problem of specifying machine-dependent microprograms within a generalized descriptive framework. The MARBLE language is, however, a complete language, whereas a schema is partially defined. In the former, machine-specific data objects are specified by constructs that indicate the binding of declared variables and procedures to hardware resources. Abstract microprograms—not oriented toward particular hosts—can also be described, in which case it becomes the compiler's responsibility to perform resource binding.

A schema (such as S*), on the other hand, must be instantiated before it can be used. In essence this means that for a given microarchitecture, valid primitive statements and operations and valid data objects must be defined (within the constraints of schema-defined rules) in order to obtain a complete language. Thus, resource binding must be done during instantiation. Finally, the schema approach is predicated on a somewhat different view of machine-independent microprogramming.

In terms of implementation, a MARBLE compiler exists; however, there has not yet been extensive application of the language to the development of real firmware. The schema S* (see next chapter) has been instantiated with respect to the Varian V75 (Varian [1975], Dasgupta [1980b]) and the Nanodata QM-1, and a partial compiler has been implemented for the latter. A fairly large S*(QM-1), program has also been written. Clearly, without additional experience no further comparative comments can be made about these approaches.

9.3 BIBLIOGRAPHIC REMARKS

The general issues and problems of high level microprogramming have also been discussed by Dasgupta (1980b), on which most of this chapter is based, Sint (1980), Davidson (1983), and Malik and Lewis (1978). Both Dasgupta and Sint have described and compared a number of different languages.

Husson (1970) provides a good survey of some of the early developments in microprogramming. It is, in addition, the most thorough and detailed reference to the microarchitecture of the IBM System/360 series.

For a discussion of the QA-1, its design, applications, and evaluation see the papers by Tomita et al. (1977), Hagiwara et al. (1980), and Shibayama et al. (1980). The QM-1 is described in detail in Nanodata (1979), while briefer descriptions are given by Agrawala and Rauscher (1976) and Salisbury (1976). An early, interesting discussion of the QM-1 in the general context of dynamic microprogramming and emulation is available in Rosin, Frieder, and Eckhouse (1972). For a description of the Varian 75 see Varian (1975), while several other user-microprogrammable computers are discussed by Agrawala and Rauscher (1976) and Salisbury (1976).

For discussions of vertical optimization of microprograms, the reader is referred to Kleir and Ramamoorthy (1971), Sitton (1973), and Tan (1978). The paper by Landskov et al. (1980) is an excellent, up-to-date and authoritative survey of the state of the art in local microcode compaction circa 1979–1980. Other important references include Ramamoorthy and Tsuchiya (1974), Dasgupta (1976, 1977), Dasgupta and Tartar (1976), Tokoro et al.

(1981), Wood (1978, 1979), Ma and Lewis (1980, 1981), Mallet (1978), Fisher (1979, 1981), Fisher et al. (1981), Rideout (1981a, b), and Yau et al. (1974).

The reference to Concurrent Pascal and its implementation are based on Brinch Hansen (1977) and Hartmann (1976).

The STRUM language has been discussed in various sources, notably Patterson (1976, 1977, 1978, 1981). The EMPL language was described by DeWitt in his Ph.D. dissertation (1976a) and is also discussed in DeWitt (1976b). A discussion of VMPL appears in (Malik and Lewis) 1978. Davidson and Shriver (1980, 1980a) describe MARBLE, while the complete language and its resource binding problems are discussed in Davidson (1980). The S* schema was developed in a series of papers by Dasgupta (1978, 1980a, 1980b).

TEN

A MICROPROGRAMMING LANGUAGE SCHEMA

The microprogramming language schema S* was originally proposed by this author in Dasgupta (1978). Substantial changes were made very soon after to the parallel constructs in Dasgupta (1980a); the review paper, Dasgupta (1980b), incorporated these changes in an informal description of S*. Based on an experimental instantiation of S* (see next chapter) further changes were made. The present description is concerned with this most recent version of S*.

10.1 DATA TYPES AND DATA OBJECTS

Table 5.1 indicated that the data types in S*A and S* are identical. Therefore I shall not repeat a detailed description of these data types but simply mention them briefly.

The only primitive data type in S* is the **bit**, consisting of the values {0,1}. Instances of type **bit** may be structured into ordered sequences or into higher-order structures such as arrays, stacks, tuples, or associative arrays.

Unlike their counterparts in S*A, all data objects are considered global. Thus, a data object declaration simply takes the form:

var x : T

which *T* names a data type. Further, unlike the S*A user, the microprogrammer writing in an instantiated version of S* cannot freely create and declare data objects as and when required; the only data objects available to the microprogrammer are those defined in the instantiated language, and these are determined according to the storage resources available in the microprogrammable host machine for which microcode is to be written. Thus as far as the user is concerned, the names, types, and structures of data objects are predefined. In an (instantiated) S* program, however, all data objects must be declared before they are referenced. The advantages of this rule are essentially threefold:

1. It forces the programmer to take explicit notice of the structure of referenced data objects, reducing the chance of errors or ambiguities.
2. The discipline contributes considerably to the clarity and readability of the program text.
3. Such declarations explicitly delineate the "scope" of the data objects, providing some measure of protection—through compile time checking—to undeclared data objects.

As is well known, variables are declared in a program for essentially similar reasons: to detect, at compile time, possibly erroneous references to undeclared variables; to enhance a program's textual clarity and enable the compiler to ensure that operations defined on variables are compatible with the declared types; and to protect variables local to a block from outside reference. This is the so-called scope rule, associated with block-structured languages.

In the case of microprograms, the effectiveness of the scope rule as a protection mechanism (3 above) is undermined by the fact that only a few data objects exist in a given host machine; hence the same data object (e.g. an array of general registers) may have to be used in different contexts in distinct parts of a microprogram—in which case the programmer may find it more convenient to declare the data objects globally. Since all data objects have to be declared globally in an (instantiated) S* program (see below for the general syntax of the S* program construct) the scope rule does not really come into effect. Nevertheless, if we were to write and compile separate parts of an emulator independently, at least some measure of protection would be afforded to undeclared variables and errors caught at compile time.

As in S*A, S* also includes synchronizers, declared in the form

$$\textbf{sync } x : T$$

where T denotes a **bit** or **sequence** data type. The **await** and **sig** primitives are defined on synchronizers in the same way as in S*A.

Example 10.1

/* data objects in the Microdata 1600 *1
var md_reg, out_reg, t_reg, u_reg : **seq**[7..0]**bit**

DATA TYPES AND DATA OBJECTS 129

```
var gpfile : array[1..30] of seq[7..0] bit
var main_mem : array[0..64K – ] of seq[7..0] bit with mar
var mar : tuple m_reg : seq [7..0] bit
                n_reg : seq [7..0] bit
         endtup
```

Example 10.2

```
/* data objects in the Varian 75 */
type register, bus : seq[15..0]bit;
type word = tuple
                 highbyte : seq[7..0]bit
                 lowbyte  : seq[7..0]bit
            endtup;
var gpr : array [0..15] of register /* gen. purp. register set */
var mar, pc, ibr, mir : register    /* sp. purp. registers */
var mem : array[0..64K – ] of word with mar, pc, ibr
```

Example 10.3

```
/* data objects in the Nanodata QM-1 */
type ls_register = seq[17..0]bit;
type f_register  = seq[5..0]bit;
var local store
   : array[0..31] of ls_register with fmod, fcod, faod, fsod, feod, gspec;
   : tuple general purpose  : array [0..23] of ls_register
           index             : array [0..3] of ls_register with fmpc
           general_purpose2  : array [0..2] of ls_register
           inst_reg          : ls_register
                             : tuple c : f_register
                                     a : f_register
                                     b : f_register
                               endtup
                             : tuple opcode : seq [6..0] bit
                                     a_parameter : seq [4..0] bit
                                     b_parameter : seq [5..0] bit
                               endtup
endtup
```

Remarks

1. In Example 10.1, the notation 64K – , used as the upper range bound for main_mem, is an abbreviation for 64K – 1.
2. All three examples illustrate the use of array with pointers. Generally speaking, an array declaration

$$\textbf{var } x : \textbf{array}[a..b] \textbf{ of } T \textbf{ with } y_1, y_2, \ldots, y_n$$

indicates that only $y_1, y_2, \ldots, y_n$ may be used as index variables for accessing x. Note that $y_1, \ldots, y_n$ must each be a **seq** type variable, or a tuple whose fields are of type **seq**. Thus, for instance, the only allowable expression involving the main_mem of Example 10.1 is

$$\text{main\_mem[mar]}$$

Similarly, the "mem" of Example 10.2 can be only accessed via "mar," "pc," and "ibr."

3. An especially useful characteristic of declarations in S* (and also S*A) is the facility to specify alternate data structures for variables (this facility is also available in the case of synonym declarations; see Chapter 5 and below). Thus, the variable "local store" in Example 10.3 is declared both as an array and as a tuple. The tuple field "local_store.inst_reg" is further defined in terms of alternate structures. In referencing "local store" the microprogrammer may reference the most appropriate representation as and when required.

As in the case of S*A, the microprogrammer may also need to declare and use predefined constants, for example

$$\textbf{const } \text{sign} : \textbf{bin}(8) \; 10000000$$

Here, the integer in parentheses gives the length of the data object in bits, and the rightmost number its fixed value.

Thus far, data objects represent entities that are visible at the microarchitectural level; their states can be changed explicitly by the action of microoperations. There may also exist in the machine entities (mainly fields or bits in microinstruction word) that, rather than effecting state transitions directly, act as gates that enable other micro-operations to effect state transitions.

Example 10.4

1. A particular set of fields in the Varian 75 microword (denoted as the TF, SF, and GF fields), when set to a specified state, enable the overflow condition to be sampled.
2. In the QM-1, the K vector contains two one-bit fields that allow lower priority and higher priority interrupts to take place at the end of the nanoinstruction's execution.
3. The K-vector also contains a one-bit field (ALU STATUS ENABLE) that allows the status of certain ALU condition flags to be stored in an

"F register" during the execution of a GATE ALU nano-operation. Similarly, the SH STATUS ENABLE field facilitates the storing of shift condition flags in an F register during the execution of a GATE SH command.

Note that none of these "enabling" operations involve storage elements other then microwords (or, in the case of the QM-1, nanowords). At an abstract level, however, these actions may be represented by assigning values to some virtual data objects. In S* such objects are termed *pseudovariables,* and may be declared in the same way as ordinary variables. For example:

pvar ovflsamp, alu_status_enable : **bit**

10.2 SYNONYMS

Although data object names are predefined in the host machine, the programmer may rename such objects using synonyms, provided these have been declared before use. The simplest form of a synonym declaration is

syn $s_1 = v_1, s_2 = v_2, \ldots, s_n = v_n$;

where $s_1, s_2, \ldots, s_n$ denote programmer-defined synonyms and $v_1, \ldots, v_n$ designate either other (previously declared) synonyms or names of data objects (with or without subscripts).

Example 10.5

var genreg : **array**[1..4] **of** register;
syn index_reg1 = genreg[1],
index_reg2 = genreg[2];

As in S*A, the programmer may also, in assigning a distinct name to a given data object, wish to associate one or more alternate structures or data types with the synonym. Such an extended synonym definition would be typically of the form:

syn S_1 : T_1
: T_2
. . .
: $T_n = V$

where S_1 is the synonym, $T_1, \ldots, T_n$ are alternate data types, and V is the name of the data object.

Example 10.6

```
var local_store : array[0..31] of seq [15..0] bit;
syn inst_reg    : tuple op1 : seq[2..0] bit
                        index : seq[1..0] bit
                        addr : seq[10..0] bit
                  endtup
                : tuple op2 : seq[2..0] bit
                        regopd1 :seq[2..0] bit
                        regopd2 :seq[2..0] bit
                        unused : seq[6..0] bit
                  endtup
                = local_store[31]
```

In this example, the data object "local_store[31]" is assigned the alternative name "inst_reg." Within the scope of the synonym declaration (see below), the data object may be viewed as an instance of three alternate data types—a **seq** of 16 bits, a **tuple** of the first form defined above, and a **tuple** of the second form. In practical terms, this allows the data object (or its constituents) to be referenced in many different ways. For example,

> local_store[31][7..0]
> inst_reg. op1
> inst_reg.addr[5..0]
> inst_reg. regopd1

The only constraint on the naming convention is that the extended names identifying a particular data object or its part should be unambiguous.

Finally, a synonym may be declared globally, or within a procedure. In the latter case, the textual scope of the declared name (and the alternatively specified data type x) is the procedure itself and any procedure nested within—unless the same synonym appears inside a nested procedure.

10.3 EXECUTIONAL STATEMENTS

10.3.1 Assignments

The basic executional entity in S* is the *assignment statement,* which is used to represent simple transfer and functional microoperations (MOs) in the host machine. Let $x_1, \ldots, x_n$ ($n \geq 1$) denotes simple or subscripted variables or pseudovariables. Also let D be a data object (i.e., a variable or a pseudovariable, a constant data object, or a literal), and E a host machine–defined arithmetic or logical expression. Then assignments are of the form

$$x_1, \ldots, x_n := D \qquad (10.1)$$

$$x_1, \ldots, x_n := E \qquad (10.2)$$

In (10.1), the value in D is transferred along some set of data paths into x_1, ..., x_n. In (10.2) the expression E is evaluated and the resulting value transferred along some set of data paths into x_1, ..., x_n. When the left-hand side of the assignment statement contains more than one variable name (i.e., $n > 1$), the right-hand value may be said to be *broadcast* to these variables. No assumption may be made as to ordering of the assignments to x_1, ..., x_n.

Note that the syntax and semantics of expressions are not specified in the definition of S*; what constitutes legal expressions, and the meanings of such expressions, are determined by the architectural characteristics of the host machine. Similarly, the legality of an assignment statement is determined not only by the legality of the right-hand expression, but also by whether the host microarchitecture permits transfers to the specified left-hand variables. This approach provides a solution to the problems posed by the variability of microarchitecture discussed in Chapter 9.

Consider now the semantics of the S* assignment. For the sake of argument I shall only consider the simpler form, involving a single left-hand variable. In general, the meaning of this statement will differ from that of the typical assignment in programming languages because of *side effects*. For example, the axiom of assignment in Pascal as given in Alagic and Arbib (1978) is of the form:

$$\{P_E^X\} \; X := E \; \{P\} \qquad (10.3)$$

This implies that (1) the evaluation of E will not change any variable, and (2) the execution of the statement may change the left-hand variable and no other. Thus, side effects are not permitted.

The assignment statement in a microprogramming language, however, reflects not merely modification of the left-hand variable but a change in the machine state in general. I have already given examples, in Chapter 9, of arithmetic operations that set condition (status) flags as side effects. As another simple example involving the QM-1, let us examine the execution of the high-level statement

$$\text{local\_store}\,[5] := \text{local\_store}\,[3] \qquad (10.4)$$

This statement involves first assigning the values 5 and 3 to the six-bit F-store registers FAOD and FAIL, and then effecting the transfer

$$\text{local\_store}\,[\text{FAOD}] := \text{local\_store}\,[\text{FAIL}]$$

That is, FAOD and FAIL are assigned values as side effects of the execution of (10.4). These side effects become part of the semantics of the statement (10.4) and cannot be ignored.

We cannot, therefore, specify a general axiom such as (10.3) for the S* assignment. The precise meanings of assignments in an instantiated version of S* will be determined by the microarchitecture, and consequently can be defined only during instantiation.

10.3.2 Timing-Related Constructs

In S*A the problem of timing and the concept of the clock cycle were deliberately ignored, since these aspects appear to be more characteristic of the register-transfer and microarchitectural levels than of exoarchitectures or even endoarchitectures. One of the basic differences between S*A and S* lies in the fact that S* includes timing-related parallel statements. Let

$$S_1 \; \theta \; S_2 ::= S_1 \; \square \; S_2 \mid S_1 ; S_2 \qquad (10.5)$$

where $S_1 \square S_2$ denotes the parallel composition of two statements while $S_1 ; S_2$ denotes the usual sequential composition of S_1 and S_2. The execution of the parallel construct $S_1 \square S_2$ causes S_1, S_2 to be executed in parallel. It terminates when both S_1 and S_2 have terminated. The S* schema includes the constructs

$$\textbf{cocycle } S_1 \; \theta \; S_2 \; \textbf{coend} \qquad (10.6)$$

$$\textbf{stcycle } S_1 \; \theta \; S_2 \; \textbf{stend} \qquad (10.7)$$

In (10.6), S_1, S_2 are simple statements or composites of simple statements; in (10.7), they are simple or composites of simple, **stcycle**, or **cocycle** statements.[1]

Given an arbitrary statement $S_1 \; \theta \; S_2 \; \theta \ldots \theta \; S_n$, a left-to-right priority of the operator θ holds unless the parentheses **do** . . . **od** or other delimiters (e.g., **stcycle** . . . **stend**, **cocycle** . . . **coend**) are encountered. In the latter cases, the operators enclosed within delimiters have the higher priority; in the case of nested delimiters the priority ordering of the θ's is analogous to that in parenthesized arithmetic expressions.

The **cocycle** construct indicates that the composite event $S_1 \; \theta \; S_2$ begins and ends in the same microcycle. The **stcycle** statement specifies that the event $S_1 \; \theta \; S_2$ begins in a new microcycle. Precisely when it will end is left unspecified, but can in principle be known from the details of S_1 and S_2. The termination of either statement transfers control to the next statement in sequence, except when the statement body specifies an explicit transfer of

---

[1] A *simple* statement is a statement in an instantiated version of S* that represents primitive microoperations in the host machine. A composite of simple statements $S_1, \ldots, S_n$ is the statement $S_1 \; \theta \; S_2 \; \theta \ldots \theta \; S_n$, possibly with the addition of parenthetic delimiters. The constructs in S* that may represent such microoperations are the assignment, the **if** statement, **call**, **return**, and the **goto** statement. However, there are restrictions on the use of the last two statements, as discussed below.

control. Thus, the basic distinction between the two constructs lies in the specified duration of their respective actions: the duration of a **cocycle** is exactly one microcycle while that of the **stcycle** is at least one, but possibly several, microcycles.[2]

Both statements satisfy the following important property: they are single-entry, single-exit constructs. That is, in each activation of a **cocycle** (or **stcycle**), every statement in its body must execute exactly once, either in parallel or in the specified sequence.

This implies that the body of a **cocycle** (or **stcycle**) is a straight line segment. Thus, it is not possible to jump (from outside) to a statement within the **cocycle** (or **stcycle**). Nor is it possible to jump from one statement to another within the body. It is possible, however, to transfer control explicitly to a statement outside the **cocycle** (**stcycle**) or to the same **cocycle** (**stcycle**) as a whole. (Note that when the body is of the sequential form, an explicit transfer of control may only occur from within the second component, S_2.)

A few examples will clarify the application of these constructs.

Example 10.7

/* Enable ALU cond.codes and perform acc := acc + gpr[y] + 1 */
   cocycle
       alusample := 1; carryin := 1; acc := acc + gpr[y]
   coend

Example 10.8

/* Read from mem and perform alu operation in parallel */
   stcycle
       ibr := mem[pc] □ acc := acc + gpr[a]
   stend

Example 10.9

   back: **cocycle**
       pctr := pctr − 1;
       if pctr ⌐= 0 => **goto** back **fi**
   coend

In Example 10.7, the three statements are executed sequentially but within the same microcycle. The two statements in Example 10.8 commence

---

[2] A microcycle is the unit of time or the duration of the clock cycle time governing the normal execution of a single microinstruction. Most microoperations execute within the microcycle; hence a microinstruction (which is, logically, a specified set of microoperations) is normally executed during one such cycle. Some microoperations (e.g. main memory read/write) may require several microcycles, in which case the execution time of the microinstruction containing such operations is correspondingly lengthened.

together, but the duration of the **stcycle** statement as a whole will depend on the timing of its components. In Example 10.9, the **cocycle** loops until pctr = 0.

Consider now the proof rules for the **cocycle** and **stcycle** statements that do not contain **return** or **goto** statements. First, the statement $S_1 ; S_2$ satisfies the standard rule of sequential composition:

$$\frac{\{P_1\} S_1 \{P_2\}, \{P_2\} S_2 \{P_3\}}{\{P_1\} S_1 ; S_2 \{P_3\}} \qquad (10.8)$$

For the parallel composition $S_1 \square S_2$, we adopt the Owicki-Gries rule (see Chapter 8, Section 8.2). To recapitulate briefly, two statements S_1, S_2 are interference-free if the execution of $S_1(S_2)$ does not interfere with the proof of $S_2(S_1)$. Thus we have:

$$\frac{\{P_1\} S_1 \{Q_1\}, \{P_2\} S_2 \{Q_2\} \text{ are interference-free}}{\{P_1 \wedge P_2\} S_1 \square S_2 \{Q_1 \wedge Q_2\}} \qquad (10.9)$$

Remark. The application of (10.9) requires that any assignment statement appearing in S_1 or S_2 must be indivisible. Thus although an instantiated version of S* could allow both high-level or primitive assignments to be written, rule (10.9) can only be applied in parallel statements involving primitive assignments.

Clearly, the above proof rules are sufficient for the **cocycle** and **stcycle** statements. That is, for the constructs **cocycle** $S_1 \square S_2$ **coend**, and **stcycle** $S_1 \square S_2$ **stend**, rule (10.9) holds, while the statements **cocycle** $S_1 ; S_2$ **coend** and **stcycle** $S_1 ; S_2$ **stend** satisfy (10.8). The timing semantics for the two classes of constructs are, however, defined informally, as previously stated.

If the **cocycle** or **stcycle** statements contain **goto** or **return** statements, the above proof rules are not applicable. However, keeping in mind the semantics of the **goto** and **return** statements (these are the same in S* as in S*A— see Appendix A, Sections 8.2.5 and 8.2.6) additional rules can be defined for a restricted set of statement forms involving the structured use of these constructs.

As a specific example, consider the statement form:

$$\begin{array}{l} \text{loop} : \textbf{cocycle} \\ \quad S_1; \\ \quad \quad \text{if } B => \textbf{goto} \text{ loop fi} \\ \textbf{coend} \end{array} \qquad (10.10)$$

This may represent the repeated execution of a single microinstruction. Following the notation introduced in Appendix A, Section 8.2.5, we may construct the following rule for the above **if** statement:

EXECUTIONAL STATEMENTS

$$\{Q\} \; \textit{if} \; B \; => \textbf{goto loop fi} \; \{R\}\{\textbf{loop}: P\}$$

This says that if Q is true initially, then on normal exit from this statement (i.e., when B is false), R must be true, while on exit through the **goto** loop, the assertion P must be true. That is, on reaching the label "loop," P must hold (P is thus an invariant associated with the label "loop"). For (10.10) we then have the following proof rule:

$$\frac{\{P\} \; S_1 \; \{Q\}, \{Q\} \; \textit{if} \; B \; => \textbf{goto loop fi} \; \{R\}\{\textbf{loop}: P\}}{\{P\} \; \textbf{cocycle} \; S_1 \; ; \; \textit{if} \; B \; => \textbf{goto loop fi coend} \; \{R\}}$$

10.3.3 Sequential Composition

In addition to the use of the **stcycle** and **cocycle** constructs, microstatements can be composed into larger sequences in essentially two ways. Conventional composition is by way of the construct:

$$\textbf{begin} \; S_1; S_2; \ldots ; S_n \; \textbf{end}$$

where S_i is any microstatement (including nested **begin** statements). Such a sequential statement may be transformed and optimized by a compiler so as to improve or parallelize the statements, so long as the implied data dependencies are preserved.

On the other hand, given the limitations of practical optimization and compaction strategies, it also seems desirable for the microprogrammer to have the choice of optimizing the program at the source level and to specify an optimal sequence. For this purpose S* provides the **region** statement:

$$\textbf{region} \; S_1; S_2; \ldots ; S_n \; \textbf{endreg}$$

where S_i is a simple, **stcycle**, or **cocycle** statement. A region denotes a segment that has presumably been optimized by the programmer. The prescribed flow of control in such segments must be preserved; hence the compiler must not perform any further optimizing transformations on such segments.

It is permissible, however, for the compiler to move code into a region body, as long as this does not violate the legality of the latter. Further, jumps into the region body are allowed as long as the rules governing the composition and execution of **stcycle** and **cocycle** statements (if present inside the region) are not violated.

10.3.4 Modularization Concepts

In S*A, a hierarchy of modularization constructs, consisting of system, mechanism, and public and private procedures, are provided in order to meet some of the requirements of architecture descriptions stated in Chapter 5. As I suggested in Chapter 1, microprograms are a means of implementing architectures, and this includes the implementation of the concepts of data abstraction, information hiding, and critical regions. Thus, the set of module-forming constructs in S* is smaller than in S*A, consisting simply of the program and the procedure constructs. All data objects are declared globally in an instantiated S* program, and the distinction between public and private procedures no longer exists.

The microprogrammer may construct procedures according to the syntax:

proc name{(par$_1$, . . . , par$_n$)}
    {<synonym declarations>}
    <procedure body>
endproc

All data objects referenced in <procedure body> are global objects. Thus, a parameter list (which is optional), if present, is redundant and serves essentially to enhance the readability of the description. Parameters, if specified, are actual parameters. The notions of formal and actual parameters and parameter passing mechanisms have no significance here. The only possible local declarations in a procedure are synonyms of global data objects.

A complete, independently compilable S* microprogram is specified by a program block:

prog name
    <declaration block>
    {<initialization block>}
    <execution block>
endprog

where <declaration block>

declaration . . . **endec**

encapsulates the data object declarations, and <execution block> is a set of one or more procedures.

The initialization block, which may possibly not exist in a given **prog** block, provides a means of initializing **seq** type data objects before execution of the program begins. Its general form is

init <init list> **endinit**

where <init list> is a sequence of one or more assignment statements.

Example 10.10

  prog sample
      declaration . . . **endec**

        init /* local store register 24 set as the microprogram counter */
            fmpc := 24
        endinit

        proc add_relative;
            fcod := data_register;
            region index_alu_out := local_store[base_register] + immediate;
                local_store [fcod] := control_store[index_adr]
            endreg;
            index [fmpc] := index[fmpc] + 2
            act fetch
        endproc
        . . . .
endprog

Notice in this example that the **act** statement is used to invoke a procedure called "fetch." Upon execution of this statement, control simply falls through to the end of the procedure and "add_relative" terminates. Note also the use of the region statement.

10.3.5 Other Control Statements

The remaining executional statements are identical to those in S*A. These include the repetition statements **while** B **do** S and **repeat** S **until** B; the two alternation (or selection) statements **if** $B_1 => S_1, \| \ldots \| B_n => S_n$ **fi** and **case** . . . **of** $V_1 : S_1; \ldots ; V_n : S_n$ **endcase;** and the **goto** construct. Since these have already been discussed in the context of S*A, the reader is referred to the relevant sections of Chapter 5.

10.4 BIBLIOGRAPHIC REMARKS

For a discussion of the influence of modern programming methodology on the design of microprogramming languages in general and on S* in particular, the reader is referred to Dasgupta (1980a).

The idea of instantiating a schema for a particular host machine is further explored by Hobson et al. (1981), who describe a language based on an APL syntax. Some parts of an instantiation of S* for the Varian 75 (S*[V75]) are described in Dasgupta (1980b). For a thorough discussion of the **goto** statement and its proof rules, the reader is referred to Alagic and Arbib (1978).

ELEVEN

A PRACTICAL MICROPROGRAMMING LANGUAGE

11.1 INTRODUCTION

The schema S* discussed in Chapter 10 is not directly usable. For it to be so we have to instantiate the schema with respect to a particular microprogrammable host machine. In this chapter I shall describe a practical microprogramming language, S*(QM-1), described in Klassen (1981) and Klassen and Dasgupta (1981), which was obtained by instantiating S* with respect to the Nanodata QM-1 (Nanodata [1979]).

As is well known, the QM-1 has two levels of control (Fig. 11.1). The higher level allows the user to write vertical, 18-bit wide microinstructions that reside in the QM-1 control store. Such control store–resident microprograms interpret instructions contained in main store. The lower level of control allows the user to write horizontal, 360-bit wide nanoinstructions, an ordered sequence of which—a nanoprogram—resides in the QM-1 nanostore and interprets microinstructions from control store.[1]

---

[1] The architecture of the QM-1 also allows main store instructions to be interpreted directly by nanocode. Klassen (1981) termed this interpretive mode *direct emulation,* in contrast to the technique of *indirect emulation* schematized in Fig. 11.1 (see also Klassen and Dasgupta [1981]). Thus, the emulator designer may exercise a choice between direct, indirect, and hybrid (or "mixed mode") emulation in implementing a target exoarchitecture on the QM-1. An interesting application of direct emulation is the emulation of the PDP-11 exoarchitecture described by Demco and Marsland (1976).

INTRODUCTION

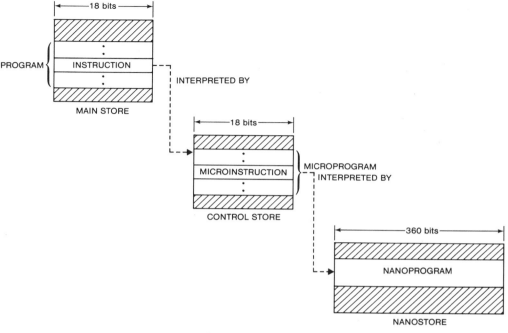

Figure 11.1 Storage hierarchy in the QM-1.

By convention, the architecture of the QM-1 at the microprogram level is termed the QM-1 microarchitecture, and that at the nanoprogram level its nanoarchitecture. The S*(QM-1) language is designed for the description of high-level nanoprograms; that is, it was obtained by instantiating S* with respect to the QM-1 nanoarchitecture.

Remarks. A note on the background to this instantiation effort: the development of S*(QM-1) was undertaken by Klassen (1981) as a research project, to examine the usefulness of S* as a schema from which a language could be developed for a real machine. The QM-1 was chosen as the host since, with its very wide nanoinstruction word and the use of residual control,[2] it afforded a rather stringent test of the adequacy of the schema. Concurrently with the instantiation effort, a study of efficient, automatic nanocode generation for the QM-1 using the linear microcode compaction algorithm was undertaken (Dasgupta and Tartar [1976], Landskov et al. [1980]), since the successful use of a high-level microprogramming language hinges rather critically on this factor. For more on the S*(QM-1) project, see

---

[2]*Residual control* is the idea of filtering relatively unchanging control information out of the microinstruction into one or more registers that may be set up by a microinstruction. The main advantage is that it helps to reduce microinstruction size. For further discussion, see Footnote 6 of this chapter.

142   A PRACTICAL MICROPROGRAMMING LANGUAGE

Section 11.6 (Bibliographic Remarks) at the end of this chapter and also Chapter 15.

The rest of this chapter is organized as follows. Section 11.2 provides an overview of the QM-1 nanolevel architecture. Since S*(QM-1) is a machine-specific instantiation of S*, some knowledge of the host architecture is essential for an understanding of the instantiated language.

Section 11.3 describes the basic structure of S*(QM-1), and Section 11.4 gives a short sample program in the language. Finally, Section 11.5 discusses the significance of S*(QM-1) in the general context of high-level microprogramming.

11.2 ARCHITECTURE OF THE QM-1

A clean description of the QM-1 nanolevel architecture is by no means a trivial task. A first glance at its data path organization (Nanodata [1979], Salisbury [1976]) (Fig. 11.2) appears to suggest that the QM-1 fails to satisfy the basic tenets of complexity management: It does not seem to be hierarchi-

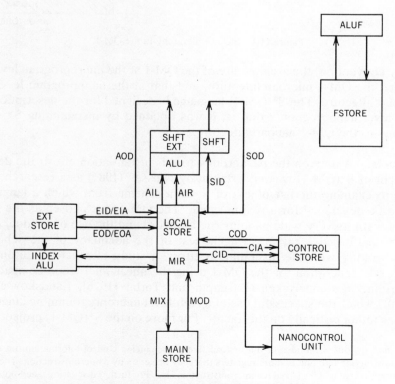

Figure 11.2 Data path organization in the QM-1.

ARCHITECTURE OF THE QM-1   143

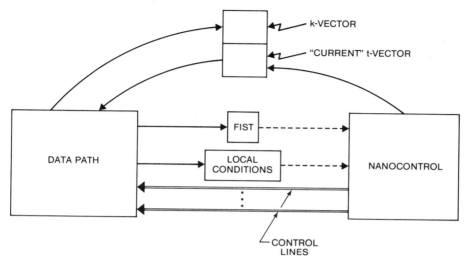

Figure 11.3 Overview of the QM-1.

cally structured, nor is there any noticeable degree of modularization; finally, its components appear to be rather strongly coupled with one another—it cannot be characterized as a nearly decomposable system.

These prima facie observations make the task of describing the QM-1 in a well-structured manner something of a challenge. I shall, accordingly, attempt to formulate a description that imposes some semblance of hierarchy, modularity, and near decomposability on what appears to be an inherently unstructured architecture.

Figure 11.3 sketches a first-level decomposition of the system. The QM-1 consists of a data path (DP) part and a nanocontrol (NC) part. The interface between these two parts is a set of control lines (which enable NC to invoke actions within DP), a set of data objects (collectively termed the k-vector and the "current" t-vector) that can be assigned values by NC and are sources of values to the DP part, and two sets of data objects (one of which collectively forms a particular register called FIST while the other forms the so-called local conditions). These can be set by actions within DP and sampled by the NC part; these two sets of data objects, in fact, hold conditions that can be tested by nanocontrol.

In S*A notation, we may represent the overall composition of the QM-1 as follows:

sys QM1;
    type kvector, tvector = **seq[71..0]bit**;
    type freg = **seq[5..0]bit**;
    privar fstore : **array[0..31] of** freg;
    privar nanoword : **tuple**

```
                    k : kvector;
                    t : array [1..4] of tvector
                                with tindex
                        endtup;
        privar tindex : seq[3..0]bit; /* tvector index */
        privar local_cond : tuple
                                carry, sign, result, ovflow : bit
                            endtup;
        syn curr_t = t[index];
        syn fist
            : tuple sh_high, carry,
                    sign, result, ovflow, sh_low : bit
              endtup = fstore [18];
        /* end of system-wide declarations */
        sys data_path; . . . . endsys;
        sys nano_control; . . . . endsys
endsys /* QM1 */
```

11.2.1 The Data Path Part

Viewed conventionally, the data path part is a collection of functional units and storage elements that are interconnected through a system of buses. It is, however, conceptually easier to regard DP as a set of weakly coupled information processing systems activated from NC. The most important of these systems are

1. The ALU system.
2. The shifter system.
3. The index ALU system.
4. The main store system.
5. The microprogram counter (MPC) system.
6. The control store system.
7. The Fstore system.

Each of these systems are described briefly below.

The ALU System

Basically, this system (Fig. 11.4) consists of a collection of arithmetic and logic functions that may be performed on 18- or 16-bit data. The inputs to these functions are values on buses AIL and AIR (left and right inputs, respectively, to the ALU), and CIH, the carry-in-hold; the results are placed on the AOD bus via an entity called the *shifter extension*.

In addition, four ALU *status conditions*—carry (C), sign (S), nonzero

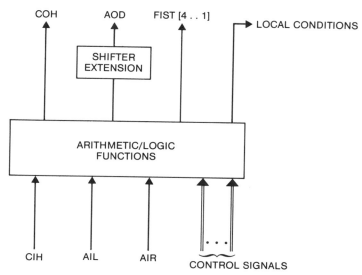

Figure 11.4 The ALU system.

result (R), and overflow (O), collectively termed "local conditions"—are produced as side effects. The carry condition may, in addition, optionally be placed in COH, the carry-out-hold flip flop, while all four conditions may optionally be stored in more permanent form in a 6-bit register called FIST.

We may therefore specify the overall composition of the ALU by the following mechanism:

```
mech ALU;
    type bus, reg = seq[17..0] bit;
    glovar fist : tuple sh_high  : bit
                       carry    : bit
                       sign     : bit
                       result   : bit
                       ovflow   : bit
                       sh_low   : bit
                 endtup
    glovar local_cond : tuple
                           carry   : bit
                           sign    : bit
                           result  : bit
                           ovflow  : bit
                        endtup;
    glovar shifter_extn : reg ;
    glovar cih, coh     : bit;
    chan ail, air, aod  : bus;
```

/* procedure definition part—one proc. for each
     distinct alu function */
....
endmech

The Shifter System

From a logical point of view, this system (Fig. 11.5) is capable of performing left and right shift functions (circular, logical, and arithmetic) on 18-, 36-, and, for certain types of shifts, 37-bit values. Inputs to 18-bit shift functions are obtained from the SID bus and their outputs are placed on the SOD bus. In the case of 36-bit shifts, the high- and low-order 18-bit inputs are obtained from the shifter extension and SID respectively, and the high- and low-order 18-bit outputs are placed respectively on AOD and SOD buses. When a 37-bit shift is involved, the 37th bit input is taken from COH—the carry-out-hold flip flop—which also serves as a destination in the case of 37-bit shifts.

The shift amount is specified by the value of a 6-bit field in the k-vector (see Fig. 11.2) called KSHA. In addition, as in the case of the ALU system, two testable conditions are produced, SHB and SLB, the high- and low-order bits of the output respectively. These can also be placed in more permanent form in the high- and low-order bits of FIST.

Since the shift functions can only be activated one at a time and they operate on the same collection of data objects, the overall system may be described as an S*A mechanism:

mech SHIFT;
   type bus, reg = seq[17..0]**bit**;
   glovar fist : **tuple** sh_high : **bit**
           . . .
               sh_low : **bit**
           endtup
   glovar shifter_extn : reg
   glovar coh, shb, slb : **bit**
   glovar ksha : seq[5..0]**bit**
   chan sid, sod, aod : bus
      . . .
   /* procedure definition part—one proc. per distinct
        shift function */
      . . .
endmech

The Index ALU System

This system (Fig. 11.6) is designed primarily to provide a fast indexing capability at the nanoarchitectural level. It is, in fact, an independent arithmetic and logic system in its own right, capable of a large repertoire of functions.

ARCHITECTURE OF THE QM-1

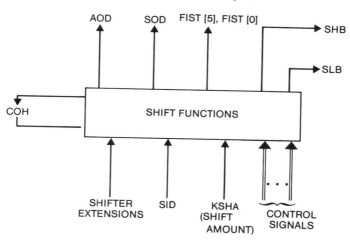

Figure 11.5 The shifter system.

The index ALU system is arguably the most complex component in the QM-1 data path part, in that an unusually large number of data objects participate in the system's functioning. This has two contributory effects on its overall complexity: first, the system's interface with its environment is not as thin as one would desire; and second, many of these data objects are used indirectly for selecting the actual inputs to the index ALU functions.

The source of the left input value to an index ALU function can be any one of 28 of the 32 local_store registers (e.g. local store [31..28], local _ store [23..0]). The right input value can be taken from one of 12 index

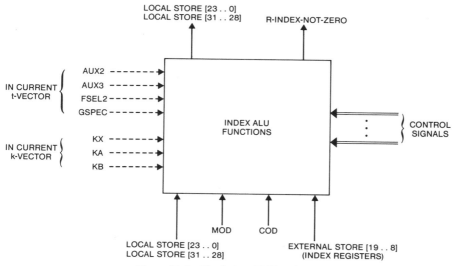

Figure 11.6 The index ALU system.

registers in external store (e.g. ext_store [19..8]), or from the MOD or COD buses. The result of an index ALU function can go to one of the same 28 local store registers; a one-bit testable value, "r_index_not_zero," is also generated.

The primary interface of this system with its environment, then, is quite thin. However, a number of additional data objects that physically reside in the current t- and k-vectors of the current nanoword must also be considered, since whenever the system is activated, the selection of the actual sources of input values and the destination of the result are determined through several levels of indirection involving these additional data objects.

The precise manner in which the input sources and the output destination are selected is best described in formal notation. The overall structure of the index ALU system may be depicted as follows:

    sys INDEX_ALU;
        mech INPUT_INDX_ALU; . . . **endmech**;
        mech OUTPUT_INDX_ALU; . . . **endmech**
    endsys

Here, INPUT_INDX_ALU is responsible for both selecting the input values and performing the arithmetic and logical functions; OUTPUT_INDX_ALU simply places the result in the appropriate destination register.

A refinement of INPUT_INDX_ALU leads to the following structure:

    mech INPUT_INDX_ALU;
        /* declaration of data types and data objects */
        . . . . .
        proc INPUT; /* decode and input values to alu */
          . . .
        endproc;
        proc INDX_ALU_FUNCT **priv**; /* perform alu functions */
          . . .
        endproc
    endmech

Of primary interest here is the INPUT procedure, which when activated, selects the appropriate input values to the index ALU hardware unit and then invokes the private procedure INDX_ALU_FUNCT; the latter performs the actual ALU functions.

Figure 11.7 is a specification of the declaration part of this mechanism, while Fig. 11.8 describes the INPUT procedure.

Remarks

    1. It can be seen that aux2 selects one of six six-bit variables that in turn determines which element of local_st should be the left input source.

ARCHITECTURE OF THE QM-1   149

```
/* declaration of data types and data objects inside
                              INPUT_INDX_ALU */
 type bus, reg = seq[17..0]bit;
 type freg      = seq[5..0]bit;
 glovar local_st, ext_st : array[0..31]of reg;
 glovar fstore : array[0..31]of freg;
 chan cod, mod : bus
 privar index_alu_left, index_alu_rt : reg;
 privar param_value : seq[5..0]bit;
 glovar index_alu_out : reg;
 /* synonym declarations */
 syn ka = k[49..44], kb = k[43..38], ksha = k[25..20],
         ks = k[19..14], kt = k[13..8], kx = k[7..2];

 syn aux2 = curr_t[11..9],
         aux3 = curr_t[8..6],
           fsel2 = curr_t[22..18];
 syn gspec = curr_t[36..33];
 syn mir = local_st[31];



 syn a = mir[11..6], b = mir[5..0];
    syn gstore = fstore[20..31], index_regs = ext_st[8..19];
    const all_ones = dec(18) 11..1;
```

        Figure 11.7 Data objects in index ALU.

When the value of aux2 is 5, an additional level of indirection is required.

2. The right input source is selected according to the contents of one of eight six-bit registers; the latter in turn is selected according to the value in aux3.

3. When the value in the six-bit register selected by aux3 is 13, the right input source is a special central processing unit (CPU) control register which, for the sake of simplicity, has not been defined in Fig. 11.7.

```
proc INPUT;
    /* decode and input values to left and right inputs of
                                index alu in parallel */
    do
       case aux2 of
         0 : index_alu_left := local_st [a]
         1 : index_alu_left := local_st [b]
         2 : index_alu_left := local_st [kx]
         3 : index_alu_left := local_st [ka]
         4 : index_alu_left := local_st [kb]
         5 : index_alu_left := local_st [gstore[gspec]]
       endcase
    ¤ do
       case aux3 of
         0 : param_value := a
         1 : param_value := b
         2 : param_value := kt
         3 : param_value := kb
         4 : param_value := gstore[8]
         5 : param_value := gstore[9]
         6 : param_value := gstore[10]
         7 : param_value := gstore[11]
       endcase;
       case param_value [3..0] of
         0 : index_alu_rt := index_regs [0]
         1 : index_alu_rt := index_regs [1]
         2 : index_alu_rt := index_regs [2]
         3 : index_alu_rt := index_regs [3]
         4 : index_alu_rt := index_regs [4]
         5 : index_alu_rt := index_regs [5]
         6 : index_alu_rt := index_regs [6]
         7 : index_alu_rt := index_regs [7]
         8 : index_alu_rt := index_regs [8]
         9 : index_alu_rt := index_regs [9]
        10 : index_alu_rt := index_regs [10]
        11 : index_alu_rt := index_regs [11]
        12 : index_alu_rt := all_ones
        13 : index_alu_rt := "control registers"
        14 : index_alu_rt := mod
        15 : index_alu_rt := cod
       endcase
    od
    od ;
    /* invoke alu funct */
    call INDX_ALU_FUNCT;
endproc
```

Figure 11.8 INPUT procedure for index ALU.

4. Notice that while fstore is defined as a private variable in the overall QM-1 system, it may well be a global variable relative to some enclosed system or mechanism. This is, in fact, the case with regard to INPUT_INDX_ALU, as can be seen in Fig. 11.7.

5. It will also be noted that "index_alu_left" and "index_alu_rt" are declared as private to the mechanism, while "index_alu_out" is defined as a global variable. This is because the latter is the interface between the INPUT_INDX_ALU and the OUTPUT_INDX_ALU mechanisms.

ARCHITECTURE OF THE QM-1 151

```
mech OUTPUT_INDX_ALu;

/* declarations of data types, data objects, and
synonyms */

.........

    proc OUTPUT;
        case gspec of
            12 : local_st [ksha] := index_alu_out
            13 : local_st [b]    := index_alu_out
            14 : local_st [ks]   := index_alu_out
            15 : local_st [kx]   := index_alu_out
            else : local_st [gstore[gspec]] := index_alu_out
        endcase
    endproc
endmech
```

Figure 11.9 OUTPUT mechanism for index ALU.

6. Finally, it must be pointed out that "param_value" has been defined only to enhance the clarity of the selection mechanism, as described in S*A. There is no indication in the QM-1 reference manual (Nanodata [1979]) that such a data object exists. In fact, the two **case** statements in the lower half of Fig. 11.8 can be quite simply implemented through a decoder complex.

The second of the INDEX_ALU mechanisms is OUTPUT_INDX_ALU. This is described in Fig. 11.9. To save space, the data object and type declarations have been omitted, since they are already shown in Fig. 11.7. Notice the much abbreviated form of the **case** statement here. For any value i of GSPEC ($0 \leq i < 12$), local_st [gstore[i]] is selected as the destination for the result. (A similar abbreviation could have been used in Fig. 11.8.) Finally, note that the OUTPUT_INDX_ALU mechanism would be activated from within INPUT_INDX_ALU. INDX_ALU_FUNCT (not shown here) after the appropriate function has been computed.

The Main Store System

The systems described so far are all instances of procedural abstractions—entities that accept a set of inputs and map them onto a set of outputs. Clearly, all hardware functional units can be thus modeled. The main store system, in contrast, can be modeled as a data abstraction—that is, as a set of

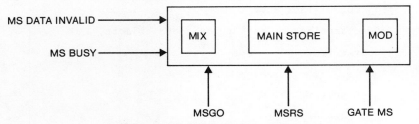

Figure 11.10 The main store system.

objects and a collection of operations defined on these objects, such that the only way of manipulating these objects are through the defined operations. The latter themselves will be procedural abstractions.[3]

The main store (MS) system (Fig. 11.10) consists of the following entities: two 18-bit buses, MIX and MOD; the main store itself, an array of 18-bit words; and a set of operations defined on these data objects, MSGO, MSRS, and GATE MS.

The maximum size of the main store is 1024K words. The MIX unit acts as both the input data bus and the memory address bus, while MOD is the memory output bus. The main store may be tested for two status conditions—whether or not it is busy and whether or not the data being accessed from main store are available, by means of the test operations MS BUSY and MS DATA INVALID respectively.

The ultimate objectives of main store operations is to *read* data from main store into MOD, and to *write* data from MIX into main store, at an address previously specified in MIX. The defined operations in the main store system are low-level means of effecting these actions.

To initiate a full (i.e. nondestructive) read operation on main store, a test (MS BUSY) must first be performed to determine that main store is not busy. If so, both MSGO and MSRS are performed simultaneously. The result is that main store is accessed at the address specified in MIX.[4] When

---

[3]Strictly speaking, procedural and data abstractions are intended to be proper abstractions. That is, in the case of the former, the algorithms by means of which the input/output mapping is realized are hidden in the specification. Similarly, in the case of a data abstraction, only the observable effects of the operations are defined, not how these effects are realized. Thus, a description of the index ALU system qua procedural abstraction should, strictly speaking, only specify the input/output behavior of the procedures; since S*A is a procedural, rather than functional, language, we are not able to enforce this discipline. Nevertheless, the terms "procedural abstraction" and "data abstraction" will be used in this discussion, since they capture very well the spirit in which I wish to describe the QM-1 system. For further discussion of these concepts the reader is referred to Liskov (1980).

[4]This is a simplification of the actual situation. The value in MIX is interpreted as a displacement that is added to the contents of a *base register* (ext_st[16]) to yield the actual address. If this address falls outside the range specified in a *field register* (ext_st[17]), then a main store address violation condition (interrupt) is generated.

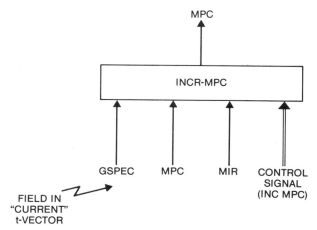

Figure 11.11 The MPC system.

the accessed word is available (signified by the MS DATA INVALID condition being false), it may be placed in MOD by means of the GATE MS operation.

To perform a write operation, an MSGO operation is first performed. This initiates the first half-cycle (the read portion) of a split-cycle operation. As in the case of a full read, the address is taken from the MIX bus. At any time thereafter, including immediately after MSGO, if the data to be written are placed on MIX and MSRS is executed, the second half of the split cycle is completed.

Finally, a read/modify/write operation may be effected by first initiating a split cycle as just mentioned. When the data become available, the GATE MS operation is then used to place the data onto MOD for modification as desired. Note that the second half of the split cycle is suspended at this stage. When the modified word is ready for insertion back into main store, it is placed on MIX and the MSRS operation is executed.

The MPC System

In essence, this system (Fig. 11.11) consists of a functional unit, shown in the figure as "incr_mpc," the microprogram counter, MPC (one of local_store [24..27] as designated by the Fstore register FMPC), MIR (local_store [31]), and GSPEC, a field in the current t-vector.

The functional unit incr_mpc continuously produces as output the values MPC + 1, MPC + 2, MPC + mir[11..0], and MPC + mir[5..0]. Under the action of a control signal (corresponding to the nanoprimitive operation INC MPC), one of these four output values—as determined by the encoding of the GSPEC field—is placed back into MPC.

The Control Store System

The control store system is shown schematically in Fig. 11.12. It consists of an 18-bit wide control store (of maximum capacity 16K words); the CIA, CID, and COD buses; the data object "index_alu_out," which holds the output of an index ALU function (see earlier discussion of the index ALU system); the microprogram counter MPC; a field "cs_addr_sel" inside the current t-vector; and the incr_mpc unit. Two operations. READ_CS, and WRITE_CS, serve to complete the system.

As in the MPC system, incr_mpc produces as output the values MPC + 1, MPC + 2, MPC + mir[11..0], and MPC + mir[5..0]. The source of a control store address, as encoded by the CS_addr_sel field, can be the CIA bus, MPC, a value in index_alu_out, or one of the values produced by incr_mpc.

The effect of a READ_CS operation is to select a control store address and read the word into the COD bus.[5] The effect of WRITE_CS is to write the value in CID (which is connected to a local store register as designated by the Fstore register FCID) into the selected word in the control store.

The Fstore System

The Fstore system (Figure 11.13) provides the primary basis for residual control in the QM-1.[6] The heart of the system is a set of 32 six-bit registers collectively termed the Fstore. Of these, Fstore [0..13] are used for specifying bus/local_store interconnections. Thus, for example, the value placed in FMOD (a synonym for Fstore [1]) determines which local_store register is connected to the MOD bus; the value in FCIA (Fstore [2]) determines which local_store register is connected to the CIA bus; and so on.

Six of the Fstore registers, Fstore [14..19], have special functions. For example, FMPC (Fstore [16]) selects which of local_store [24..27] is to be used as the microprogram counter (MPC). The FIST register (Fstore [18]) holds global status conditions resulting from ALU and shift operations (see discussions above of the ALU and shifter systems).

Finally, the registers in Fstore [20..31] are also referred to as G registers (G[0..11]); these provide service as general purpose six-bit registers that

---

[5] The value read into COD can be gated into a local store register (as specified by the Fstore register FCOD) by executing the nanooperation GATE_CS.

[6] The idea of "residual control," developed by Flynn and Rosin (1971), is essentially the notion that in using microprograms for emulating a target architecture, a part of the control information, once set up, will remain unchanged during the course of the emulation. That is, a portion of the control information will be relatively invariant for significant periods of time. Instead of holding this information in the microinstruction (or, in the case of the QM-1, the nanoinstruction), one can place it in special registers and hold it there for any desired period of time. By reducing the amount of control information that needs to be held in the microinstruction (nanoinstruction), one can significantly reduce the width of the latter. The state of these special residual control registers can of course be altered if necessary under microprogram control.

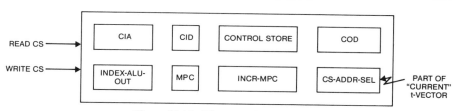

Figure 11.12 The control store system.

may be used for temporary storage, as scratchpad registers, and as backup for the other Fstore elements.

Transfer of information to and from Fstore is effected by using a compendium of other data objects, collectively termed AUX. In particular, AUX includes the three six-bit fields in local_store [31] (which, we recall, is the microinstruction register, MIR), and several six-bit fields (KA, KB, KALC, KSHC, KSHA, KS, KT, and KX) in the k-vector. In addition, the G registers can also serve as source of data for transfer to the other Fstore registers.

During a transfer to or from Fstore, the selection of AUX and Fstore registers is specified through special fields (viz. AUX0, AUX1, AUX2 for the AUX field, FSEL0, FSEL1, FSEL2 for Fstore registers) in the current t-vector. Control signals (IN0, ..., IN2, OUT1, ..., OUT3) from the current t-vector determine whether the Fstore registers selected are to be destinations or sources of the transfers. Finally, note that up to three transfers involving the Fstore registers can take place simultaneously.

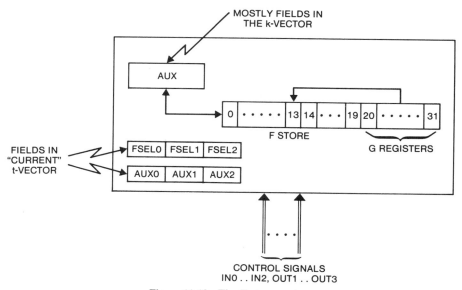

Figure 11.13 The F store system.

11.2.2 Nanocontrol

As shown in Figure 11.13, the k-vector and the current t-vector constitute part of the interface between data path and nanocontrol. Indeed these components of the interface can legitimately be viewed as the latter's contribution to the interface, in contrast to FIST and "local conditions," which are contributed by the data path part. We now turn our attention to the architecture of nanocontrol.

Consider the archetypal control unit in a microprogrammed computer (Fig. 11.14). It consists of a control store—an array of microwords—in which microprograms reside as a linear sequence of microinstructions, a microinstruction register, a control store address register (CAR), and some amount of sequencing and decoding logic. The execution of a microprogram involves fetching a microinstruction from control store at the address specified in CAR into MIR, decoding the microinstruction, and issuing control signals to the data path part of the processor. At the same time the CAR is incremented to identify the next sequential microinstruction; the microinstruction fetch/decode/execute cycle would then be repeated.

If a branch is executed from the current microinstruction, then a change in the flow of control is effected by placing the branch_to address (which would usually appear as a special field in the current microinstruction) in the CAR instead of the normal incremented value.

The microinstruction fetch/decode/execute cycle is normally completed within a fixed time, called the *microcycle*. Some microinstructions, however, may require several microcycles, for example those involving main store read/write.

The QM-1 nanocontrol is a somewhat complicated variant of this archetype (Fig. 11.15). Here nanoprograms reside in nanostore, consisting of linear sequences of 360-bit nanowords. Each nanoword consists of a 72-bit k-vector and four 72-bit t-vectors designated, respectively, T1, T2, T3, and T4. Nanostore is available in 256-word blocks up to a maximum of 1024 words. Blocks are further segmented into 128-word pages.

Mechanisms are provided for selecting a nanoword, fetching it from nanostore, and loading it into the control matrix (which is, then, the functional equivalent of the microinstruction register of Fig. 11.13). As soon as this is done, T1 immediately becomes the active or current t-vector; the k-vector also becomes active.

During normal operation of the control matrix, the four t-vectors are activated in succession and in a circular fashion: T1, T2, T3, T4, T1, and so on. Unless a program check interrupt intervenes, this sequence continues unless certain program control nano-operations are executed from within one of the t-vectors. Notice that in the QM-1, a nanoinstruction (as a functional equivalent of the archetypal microinstruction) is composed of the k-vector and the current t-vector within the control matrix. Essentially, then, a nanoword represents up to four nanoinstructions.

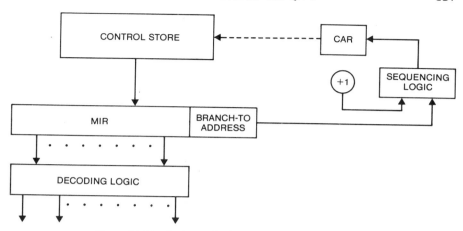

Figure 11.14 Archetypal microprogrammed control unit.

At this stage, it is important to note the following timing aspects of nanocontrol. A t-step designates a single step of nanoprogram execution, and generally speaking will consists of the simultaneous (parallel) execution of some set of nanooperations issued from a single t-vector. The duration of a t-step will normally be 80 nanoseconds (also referred to as a t-period), although for certain purposes this may be expanded (under program control, through execution of the STRETCH nano-operation) to last for two t-periods.

All nanooperations are classified as either leading edge or trailing edge operations, according to whether the functions they define take effect at the beginning or at the end, respectively, of the t-step in which they are executed.

The distinction between a t-step and a t-period becomes important when both leading and trailing edge primitives are programmed. For example if READ CS and GATE CS (leading and trailing edge operations, respectively) are specified in the same t-step, and the t-step is expanded by a STRETCH operation, the value gated into local_store will be that generated by the GATE CS from control_store, since the time span between leading and

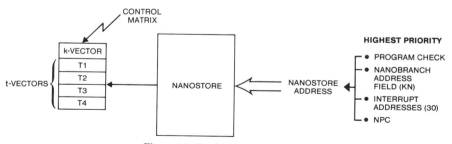

Figure 11.15 QM-1 nanocontrol.

trailing edges of a stretched t-step is two t-periods—sufficient for a control_store cycle. On the other hand, had STRETCH not been specified, the control_store will not output the new value onto the COD bus in time for the GATE CS operation to gate it into local_store.

Two nanooperations of particular interest are SKIP and GATE NS. Either can be executed conditionally, according to the t-vector test facilities. The presence of SKIP in a t-vector causes the following t-step to be skipped—that is, activation of the following t-vector is inhibited. This is, effectively, a NOOP operation of duration one t-period. Note that a SKIP executed in T3 results in T1 being the next t-vector being activated; a SKIP in T4 transfers control to T2.

The GATE NS nanooperation is a trailing edge nanooperation that causes the control matrix to be loaded with the nanoword resulting from the most recently completed nanostore access. Control is then transferred to the first t-vector of the newly loaded nanoword.

Nanostore access is initiated by means of a READ NS operation; the read operation is completed within two t-periods. Since READ NS and GATE NS are, respectively, leading and trailing edge operations, a successful nanoword fetch may be completed by either of the following sequences:

1. t-period T_n : READ NS (not stretched)
   t-period T_{n+1} : GATE NS (not stretched)
2. t-period T_n : STRETCH, READ NS, GATE NS.

11.2.3 Nanostore Addressing and Branching

When READ NS is executed, a priority select mechanism provides the actual nanostore address from a list of potential addresses (Fig. 11.15). Each source of potential nanostore address has a fixed priority relative to the other sources. On activation, the priority select mechanism selects the address from the highest priority source (whose status is ACTIVE).

As indicated in Fig. 11.5, the nanobranch address has a high priority. The address source is a 10-bit field, called KN within the k-vector, and a branch is specified by another field in the k-vector. When this field is set, and no program checks exist, the address for the next nanoword is taken from the KN field.

When a nanobranch is not specified or does not take place, and there are no interrupts pending, the priority select mechanism takes the value in the nanoprogram counter (NPC) as the address. The NPC can be loaded from different sources or simply incremented under nanoprogram control.

11.3 DESCRIPTION OF S*(QM-1)

Armed with this overview of the QM-1 architecture, we now turn to the main subject of this chapter: the design of the microprogramming language

S*(QM-1). This language, it will be recalled from Section 11.1, is an instantiation of S* for the nanolevel architecture of the QM-1. That is, programs written in S*(QM-1) are high-level representations of QM-1 nanoprograms. Much of this discussion is based on Klassen's thesis, as reported in Klassen (1981) and Klassen and Dasgupta (1981).

The general structure of an S*(QM-1) program is of the form:

prog ("program name")
   declaration
     . . . . . .
   endec
   init
     . . . . .
   endinit
   proc "procedure name" ("procedure parameters")
     . . . . .
   endproc
   "additional procedures"
endprog

11.3.1 Declarations

The main component of a *declaration block* would be data type and data object declarations whose scope extends over the entire program. That is, all data object declarations are global in their scope.[7] As defined in the parent schema S*, the primitive data type is the **bit,** while structured data types in S*(QM-1) can be **sequences, arrays,** or **tuples.** The stack and associative array are absent in this language, since corresponding hardware entities are nonexistent in the QM-1.

As in S*, data objects in S*(QM-1) can be variables or constants. The former, in turn, may be real variables or pseudovariables—the latter representing QM-1 entities that are not, strictly speaking, data objects, but are testable status flags, or fields in the k- and t-vectors that, from an abstract viewpoint, may be regarded as variables.

Example 11.1

   type ls_register = **seq**[17..0]**bit**
   type f_register = **seq**[5..0]**bit**
   . . . . .
   var local_store
     : **array** [0..31] **of** ls_register **with**
          fmod, fcod, faod, fsod, feod, gspec

---

[7]For remarks on the notion of the *scope rule* of variable declarations in the context of microprogramming languages, see Section 10.1.

```
            : tuple
                general_purpose : array [0..23] of ls_register
                index : array [0..3] of ls_register with fmpc
                general_purpose2 : array [0..2] of ls_register
                instruction_reg : ls_register
                        : tuple
                                c : f_register
                                a : f_register
                                b : f_register
                        endtup
                        : tuple
                                opcode : seq[6..0]bit
                                a_par : seq[4..0]bit
                                b_par : seq[5..0]bit
                        endtup
        endtup
```

Since the data objects in S*(QM-1) are machine-specific, their names and structures are defined during instantiation. The programmer may, however, rename declared data objects using the synonym construct:

syn "synonym name" = "variable_id"

where "variable_id" identifies a simple or structured data object.

Example 11.2

syn pdp11_register = local_store[0..7]

Remarks

1. Note that in S*(QM-1), unlike in S*A, synonyms may not be defined with alternative structures, since that would imply some direct capability of the underlying nanoarchitecture to manipulate such arbitrarily defined structures. The only kinds of structures for a given data object that the nanoarchitecture supports are those determined and defined during instantiation.
2. It is assumed that the programmer is responsible for his or her own array bound checking, in the case of a synonym declaration such as Example 11.2.
3. The main data paths in the QM-1 are all defined as variables of type "bus," the latter being defined as

type bus = **seq**[17..0]**bit**

Thus, for example, S*(QM-1) includes, among others, the predefined declarations:

    var main_store_output, control_store_output, alu_out : bus

4. As Example 11.1 illustrates, the variable local_store is declared as both an **array** and a **tuple**. The latter definition indicates that the subarray local_store [24..27] (specified as a field named "index" inside the **tuple**) is associated with the Fstore component fmpc—that is, they serve as possible microprogram counters. Alternative structures and names of local_store [31] (the microinstruction register in the QM-1) are also defined.

11.3.2 The Initialization Block

The initialization construct in S*(QM-1) allows the programmer to assign values to data objects before the start of emulator execution. Specification of which of local_store[24..27] is to be used as the microprogram counter, or assignment of values to Fstore registers to establish connections of buses to local_store elements, are typical instances of initialization statements.

Example 11.3

init
    fmpc := 25   /* establish mpc */
    local_store [25] := 100
                    /* start address of microprogram */
endinit

11.3.3 Procedures

It will be recalled from Section 11.2.3 that sequencing between nanowords is accomplished by means of a priority select mechanism that may choose, as a nanostore address source, the branch address in the current nanoword, addresses corresponding to 30 different interrupts, or the value in the nanoprogram counter. Each interrupt may be individually enabled and has the address of its first nanoword stored in an external store (ext_store) register. The nanoprogram counter may be simply incremented, loaded from control_store_output, or from the branch address in the current nanoword.

    A microinstruction (in control_store) may be invoked by loading NPC with the opcode of the microinstruction, and a three-bit *page pointer* held in an Fstore register. The opcode points to one of 128 words in one of the eight pages defined by the page pointer bits under program control.

    In essence, then, a nanoprogram segment may act as either a microinstruction interpreter, as an interrupt handler, or as a subroutine that may be

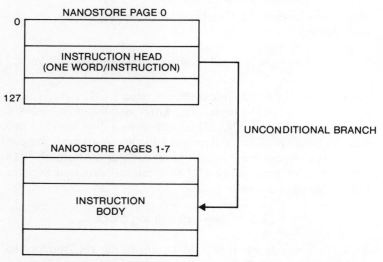

Figure 11.16 Nanostore layout for S*(QM-1) instruction procedure.

called by either instruction or interrupt routines. Accordingly, S*(QM-1) contains three specific types of procedure constructs. In each instance, the compiler will generate code to transfer control from one nanoword to the next and perform other necessary auxiliary functions.

Each procedure definition has one or more associated parameters that provide specific information to both the compiler and the reader.[8] For example, we may define a (micro)instruction procedure in the following manner:

    proc add_immediate (**instruction,** op = addi, fmt = r.r.c)
    . . . . .
    endproc

where "r.r.c." indicates that the operand format of this instruction consists of two local_store registers and an 18-bit constant immediately following the constant.

Although the S*(QM-1) programmer should not need to know about the mechanics of the compiler, it is instructive to see, for our present purposes, how the compiler would handle an instruction procedure.

Figure 11.16 shows the nanostore layout for instruction procedures. The first nanoword (the *instruction head*) is placed in a header page defined and set by the compiler. The subsequent words of the procedure—known as the

---

[8]Note that the procedure schema in S* does not provide this feature. In S*(QM-1) it serves as a very useful, but machine-specific, detail; however, further experience with instantiations for other machines may well indicate its general utility.

instruction *body*—are placed in a separate page. The head of each instruction will have a bit set that indicates that this is a legal entry point for invocation of a microinstruction. If an illegal instruction is fetched from control store, it will point to a nanoword in the header page that is not a legal entry point, thereby generating a program interrupt.

The instruction head will specify an unconditional branch from the header page to the first word of the instruction body located in one of the remaining pages, allocated by the compiler according to some convenient policy (e.g., first come, first served). The nanoprogram counter is loaded with the address of this next nanoword from the address field of the instruction head. Control within straight line nanocode segments is effected by sequential increments of the nanoprogram counter.

Interrupts are permitted to occur at the end of an instruction procedure by setting the permission bits in the last nanoword of the procedure.

The **call** statement, **call** "name" (where "name" identifies a subroutine procedure), may only be used within an instruction procedure. It invokes a subroutine, with the implication that a return will be made to the statement following the call. The return may be specified explicitly by means of the **return** statement or implicitly by reaching the end of the subroutine. The *activate* statement **act** "name" is a nonreturning call and is essentially a branch from inside a procedure to the beginning of a subroutine or some other instruction procedure. It will usually occur as the last statement in the activating procedure.

Subroutines are defined in the form:

>**proc** "procedure name" (**subroutine,** allow_interrupts)
>. . .
>**endproc**

where the (optional) parameter "allow_interrupts" indicates to the compiler that upon completion of the subroutine, interrupts may be taken.

The QM-1 supports only one subroutine level with the return address being held in NPC, which must, consequently, remain unaltered if the subroutine wishes to return control to the calling procedure. The compiler generates the flow of control within the subroutine by setting the branch address field in each nanoword to the address of the next nanoword and setting the corresponding branch bit.

Finally, interrupt procedures are defined in the form:

>**proc** "procedure name" (**interrupt,** level = . . .)
>. . .
>**endproc**

where the second parameter refers to the interrupts the procedure is to handle. Interrupts are permitted to occur on completion of a microinstruction after NPC has been loaded with the address of the next instruction. Only branch addressing is employed by the compiler within interrupt procedures, thereby preserving NPC for activation of the next instruction.

11.3.4 The Conditional Statement

The conditional statement in S*(QM-1) is of the form:

if (test condition) S **fi**

where "test condition" is one of a predefined set of machine-specific conditions and the body, S, is executed if "test condition" is true. This construct is thus a restricted form of the **if** statement in S*.

Typical conditions that may be tested are the local conditions generated from the ALU and shifter systems, **carry**, **sign**, **result**, **overflow**, **shb** and **slb** (the latter two representing the highest- and lowest-order bits in shifter_output); these same conditions saved globally in f_store.fist; and such special conditions as **ms busy** and **ms data**. An optional test modifier—**local**, **global**, or **special**—is used to specify which set of conditions is to be tested.

In addition to the above, "test condition" may be a relational expression involving two local_store variables or a test of equality or nonequality of an Fstore variable against 0.

The intrusion of QM-1-related factors in the semantics of S*(QM-1) is particularly seen in the definition of the conditional statement. For example, several restrictions are placed on the nature of the body S, depending on what type of procedure the **if** statement appears in:

1. Within **instruction** procedures, S may be any valid sequence of more primitive statements. However, if the first statement is a **cocycle**, it is assumed to be the only statement in S. If several statements beginning with a **cocycle** are required, they must be enclosed in a **begin** ... **end** block.

2. In **subroutine** and **interrupt** procedures, S may consist only of a **return** or a **cocycle** statement.

The reasons for both these restrictions are related to implementation. A **cocycle** is encoded in a single t-vector. When the body of an **if** statement is a **cocycle**, the compiler generates a **skip** nano-operation that skips over the t-vector with the **cocycle** statement. Hence the first restriction.

In the case of the second restriction above, a subroutine or an interrupt procedure, as previously noted, must not alter NPC if a successful return to the calling procedure (in the first instance) or to the next instruction (in the second) is to be achieved. Thus, the body of an **if** statement appearing in a subroutine or an interrupt procedure may not use NPC for sequencing. The only statements that may appear within such an **if** body are a **return** statement, which causes control to return to the calling procedure, or a **cocycle**, which generates a skip nano-operation.

Example 11.4

(1)   **if (local sign = 1)**
      reg[dest] := **notr** reg[source]
  fi
(2)   **if (global result = 0)**
      act jumpr
  fi

11.3.5 Iteration

Both forms of the iterative statement are present in S*(QM-1):

> **repeat** S **until** ("test condition")
> **while** ("test condition") **do** S **endwhile**

These constructs carry their usual meanings. However, a **repeat** statement that is otherwise correct in both syntax and semantics is only legal if the body S can be mapped into a single nanoword. Again, this additional restriction is necessitated by implementation factors.

11.3.6 The Case Statement

The **case** statement in S*(QM-1) differs quite fundamentally from its usual meaning as defined in, say, S*. Its general form is:

case "case name" **of**
    0 : "branch statement 0"
    1 : "branch statement 1"
    . . . .
    n : "branch statement n"
    {x := y}
endcase

where "case name" is the *name* of the **case** statement, each "branch statement" is either a **goto**, a **call**, or an **act** statement (see Section 11.3.2), and the transfer statement x := y is optional. If present, the destination of the transfer is always local_store [31]. Note that the labels of the branch statements are consecutively numbered integers, always beginning with 0.

The implementation of the **case** statement is quite intricate. Because of this, the program must have within its declaration block a **case** declaration of the form:

case "case name" : "integer"

where "integer" specifies the number of distinct clauses in the body of the **case** statement. This declaration allows the compiler to select a *base address* and reserve the required number of nanowords. Each word serves a similar function to that of an instruction header word, causing a branch to another section of a nanostore where the associated routine may be found. The labels 0, 1, . . . , n then act as offsets relative to this base address.

Only subroutine names and labels may appear within the branch statements. The fact that particular subroutines and labels are to be associated with a given **case** statement is recorded by a special parameter:

case = "case name"

appearing in the parameter section of subroutine and label declarations (see Section 11.3.8). As the compiler detects these parameters, it puts the header words of the corresponding routines in the next set of available positions in the **case** block, constructs branches, and stores the rest of the routines in other convenient areas of nanostore.

The execution of a **case** statement requires selection of one of the branch statements, depending on an offset value. This offset is placed by means of a prior assignment statement in the b field of local_store [31]—that is, in local_store.instruction_reg.b. The base address is supplied by the compiler and the offset is moved to local_store.instruction_reg.a, thereby completing the 10-bit nanostore address in the low-order 10 bits of the "ca" segment of instruction_reg (Example 11.1). The compiler supplies an illegal byte address in the b field and, on the execution of a "write ns" operation, the word in the specified nanostore address is read from nanostore.[9]

Finally, the optional assignment statement inside the **case** block allows the programmer to assign a value to local_store.instruction_reg after its contents have been used for the address but before the word read is placed in the control matrix.

Example 11.5

/* a *case* declaration in the declaration block */
case jump_table : 4

---

[9] Thus the planned implementation of the **case** statement takes advantage of a quirk in the QM-1, namely that if a 10-bit address is specified in ca fields of local_store[31] and an illegal byte address (>18) is specified in the b field, then a "write ns" operation causes a *read* to be done. For further details the reader may refer to Nanodata (1979).

```
.....
/* program segment */
local_store.instruction_reg.b := 3;
faod := 31;
case jump_table of
    0 : goto transfer1
    1 : goto transfer2
    2 : goto transfer3
    3 : goto transfer4
    local_store [faod] := alu_output
endcase
.....
label : transfer4 (jump_table) S1;
```

The execution of this program segment will cause control to be transferred to the statement S1 (whose label is transfer4), with local_store [31] containing a value assigned from alu_output. Note the label parameter in parentheses, which signifies that this particular label is associated with the **case** statement "jump_table."

11.3.7 Sequential Composition

Following S*, sequential composition is achieved in S*(QM-1) programs by means of the **begin** and **region** statements:

begin S_1; S_2; . . . , S_n **end**
region S_1; S_2; . . . ; S_n **endreg**

The **begin** construct carries with it the usual meaning that $S_1, S_2, \ldots, S_n$ are to be executed sequentially. However, if the compaction phase of the compiler (see Section 11.6) can produce a more optimal, but functionally equivalent, sequence of nanowords than may be implied by the **begin** statement itself, then the compiler is permitted to do so. In other words, it is not necessarily true that the sequential ordering of statements in a **begin** statement is preserved within the object code.

The semantics of the **region** statement at the source language level is exactly identical. However, the construct also indicates that the specified sequential ordering of its components *must* be preserved at the nanocode level.[10]

Example 11.6

> **begin** S_1;
> S_2;

---

[10]The necessity of this construct is explained in Chapter 10, Section 10.3.

```
        cocycle S₃ □ S₄ □ S₅ coend;
        S₆;
        region S₇;
            S₈;
            cocycle S₉ □ S₁₀ coend;
            S₁₁
        endreg
        S₁₂
end
```

11.3.8 Goto and Label Statements

The statement **goto** L transfers control to the statement following a **label** statement identified by the label L. Transfers may only occur to a **label** statement within the same procedure as the **goto** statement. The form of the **label** statement is

> **label** : L ("case name")

where L is the label identifier and (the optional) "case name," if present, identifies a case statement with which this label is associated. Example 11.5 illustrates the use of both **goto** and **label** statements.

11.3.9 Assignment Statements

It will be recalled from Chapter 10 that the syntax, semantics, and pragmatics of the assignment statement are only partially specified in S*. The actual legal forms of the construct become a function of the host machine architecture, and are determined during instantiation.

The machine-specific component of S*(QM-1) is most prominent in the instantiated form of this construct. While its general form is conventional, namely:

$$X_1, X_2, \ldots X_n := E$$

a variety of rules govern which statements are actually recognized within S*(QM-1).

Example 11.7

Consider simple transfer statements of the form $X := Y$, where Y is a variable identifier or a constant, and X a variable identifier. Then, representative rules are:

1. If Y is a constant it must be less than 64 and may only be assigned to

variables of types "k_vector_register" or "f_register." Pseudovariables (**pvar**'s) may only be assigned values in the range {0, 1, 2, 3}.[11]
2. Pseudovariables can only appear on the left-hand side of the assignment operator, except for the **pvar** "sw," which can only appear on the right-hand side.
3. Eighteen-bit transfers must include at least one reference to either a local_store or an external_store variable.

Example 11.8

Consider function statements of the form X := E, where E is an expression involving arithmetic, logical, or shift operators. Then the following typical rules apply:

1. Any such expression E is valid only if it consists of at most two nonoperator terms, and the specified operation can be performed in one pass through the ALU or shifter.
2. Only a subset of the total set of ALU operations are directly supported by the language. These are the unary operators **passl** (pass left), **passr** (pass right), **notl, notr, incl,** and **incr,** and the binary operators +, −, **and, nand, or, not,** and **xor**. These operators allow relatively high-level representation of expressions. For instance:

    local_store [10] := **notr** local_store [fair]
    local_store [10] := local store [5] **nand** local_store [6]

    However, if a nonsupported operation is required, the variable "kalc" must be explicitly assigned the code for the operator and a low-level form of the assignment used. For example:

    kalc := *bin* 011101; /*operation "left and not right" */
    local_store [5] :=
        local_store [10] (kalc) local_store [11];

3. The basic format of operations involving MPC (microprogram counter) is:

    $$X := X + c$$

    where X refers to MPC and c is one of the symbols 1, 2, b, and ab. For example:

    local_store.index[3] := local_store [3] + ab

---
[11]There are, in fact, further restrictions as to which subsets of these values can be assigned to specific pseudovariables.

11.4 A PROGRAMMING EXAMPLE IN S*(QM-1)

The following example describes an instruction procedure that adds the contents of the local_store register specified by the first parameter in the instruction to the immediate value following it (assumed to have already been fetched from control_store and residing on the control store output bus). This sum is then used as an address into control store. The resulting 18-bit value is placed in the local_store register (using f_store register fcod as a pointer) specified by the second parameter. The microprogram counter, as defined by the value in the f_store register fmpc, is then incremented by two and a fetch subroutine is activated.

Remark. This procedure is, in fact, emulating the PDP-11 "add relative" instruction on the QM-1.

The complete program segment is as follows:

```
prog (test)
    declaration
        /* predefined, global declarations */
        . . . . .
        syn immediate = control_store_output
        syn index_adr = index_alu_output
        syn base_register = a_par
        syn data_register = b_par

        macro INCR_MPC_2
            index[fmpc] := index[fmpc] + 2
        endmacro
    endec
    init
        /*
           local store register 24 set
           as the microprogram counter
        */
        fmpc := 24
    endinit
    proc add_relative(instruction, op=arel, fmt=r.r.c)
        fcod := data_register;
        region
            index_alu_output := local_store[base_register]
                                + immediate;
            local_store[fcod] := control_store[index_adr];
        endreg
        INCR_MPC_2;
        act fetch;
```

endproc
...
...
endprog

11.5 THE SIGNIFICANCE OF S*(QM-1)

The Nanodata QM-1, in its role as a "universal host machine," serves primarily as an engine for the emulation and evaluation of exoarchitectures, and it is in this role that it is mostly used in various architecture laboratories in North America. Thus, S*(QM-1) is unlikely to be used as a tool in a production environment. The question may then be legitimately raised as to the significance of this language design and implementation effort.

As remarked in Section 11.1, the development of S*(QM-1) served mainly as an experiment, to critically test the usefulness and expressive adequacy of S* as a linguistic framework for constructing practical high-level microprogramming languages. It was also intended to test the philosophy of instantiation as a basis for practical design.

In the light of these goals, it seems to me that S* has withstood the test to a reasonable extent. Almost all of the language S*(QM-1) was developed within the S* paradigm. In particular, the data type and data structuring capabilities, the synonym construct, and the constructs for parallel and sequential composition, as originally defined in S*, proved to be both adequate and extremely useful for both design and implementation. The few extensions or modifications made, for example in the **case** statement, or the separation of procedures into three types, reflect highly idiosyncratic properties of the QM-1 rather than weaknesses in S*.

In Dasgupta (1978) (see also Chapter 9, Section 9.2 of this work) it was suggested that, in the context of microprogramming, the notion of a "high-level" language must be revised and a somewhat more modest set of criteria established. Interpreting these objectives with respect to S*(QM-1), one would hope that the language would satisfy the following criteria:

1. A programmer using S*(QM-1) must have the capability of writing well-structured yet efficient programs that could exploit the inherent parallelism of the QM-1 at the nanoarchitectural level.
2. S*(QM-1) programs should be readable and understandable independently of the reader's knowledge of nanostore organization and its associated sequencing logic.

These criteria seem to have been met quite satisfactorily by S*(QM-1). Well-structured, readable programs that are independent of the nanocontrol structure can be written. In fact, S*(QM-1) provides essentially an abstraction of the QM-1, which frees the programmer from concerns of resource conflicts, timing, and data dependencies except as allowed by the **cocycle** and **region** constructs.

The design effort also revealed very clearly the delicate path that must be followed between the Scylla of linguistic clarity and the Charybdis of microcode efficiency. The outcome is seen quite clearly in the language structure, its syntax, and its semantics. Each construct was designed with the knowledge of precisely how it would be translated by the compiler into nanocode; in some cases the structure of the QM-1 informs certain design decisions, as in the **case** statement. In others, concerns of nanocode compaction influenced the design. Each design decision, however, was constrained by the framework imposed by S* and the objectives of high-level nanoprogramming.

The outcome has both benefits and costs. On the one hand, as the discussion in the preceding section indicates, the language definition to a large extent embeds the pragmatic aspects of translation. The designer/instantiator, like the reader, gains a high degree of confidence that S*(QM-1) can be implemented to produce reasonably efficient code. On the other hand, one cannot avoid misgivings as to how one may produce proof rules to prove S*(QM-1) programs correct or use assertions to guide the program development process. This issue remains to be studied.

11.6 BIBLIOGRAPHIC AND OTHER REMARKS

This chapter is based heavily on Klassen's (1981) thesis, which gives many more details on the instantiation of S* to S*(QM-1). A short description of this work is also reported in Klassen and Dasgupta (1981).

Klassen (1981) also discusses the design of the compiler for S*(QM-1). While a fully working compiler has yet to be completed, parts of such a compiler, namely the parser, lexical analyzer, and compactor, have been implemented.

The most critical part of the compiler is undoubtedly the compactor, which is responsible for detecting parallelism between nanooperations and packing these operations into the smallest number of nanowords. As noted in Chapter 9, the theory of microcode compaction is very well developed, especially in the case of straight line microcode (the problem of local microcode compaction). Landskov et al. (1980) is an excellent and thorough review of this topic, while additional works not reviewed there are given in Ma and Lewis (1980, 1981) and in the special issue of The Institute of Electrical and Electronics Engineers' *Transactions on Computers* (1981).

Using the theoretical model described by Landskov et al. and the so-called *linear algorithm*—an extended and modified version of a method proposed by Dasgupta and Tartar (1976)—Rideout has investigated the feasibility of nanocode compaction for the QM-1 (1981a, 1981b). An implementation of the linear algorithm was completed, and a series of tests run on intermediate code produced from S*(QM-1) programs showed that nanocode compaction is indeed feasible. The output of the compactor produced code that was consistently within 25% of hand-generated code in terms of execution time. For further discussion of the compaction problem, the reader is referred to Chapter 9, Sections 9.1 and 9.3.

TWELVE

ON STYLE IN COMPUTER ARCHITECTURE

12.1 INTRODUCTION

We do not normally think of style in the context of computer architecture. And yet wherever there is design, the problem of style arises.

A style, for our present purposes, is a set of characteristics, features, or attributes that may set one group of artifacts apart from another—artifacts that otherwise are functionally equivalent (or at least serve some common purpose). Thus, in the design of an artifact, style represents one or more acts of choice on the part of the designer, the outcome of which is the presence of one particular set of characteristics in the artifact rather than another.

As Simon (1975) has so perceptively pointed out, where there is only one way of designing a product, the problem of style disappears. Form would follow function and the resulting artifact would always possess the same set of characteristics—it would have, in other words, a unique style determined by its unique and optimal form.

Design problems do not in general have unique optimal solutions, and the designer therefore seeks *satisficing* rather than optimal solutions. Since there may be several ways of arriving at a solution that satisfices, each such design may in fact embody a particular set of design decisions that collectively imbue the product with a particular style.

Notice that style is entirely an aspect of the artificial—that which is designed, synthesized, or created—rather than of the natural. The biologist does not, for instance, in ordinary terms, differentiate between species in

INTRODUCTION

terms of stylistic differences; nor does the chemist speak of distinctive chemical structures as an aspect of chemical styles.

However, even within the realm of the artificial the notion of style is traditionally associated with products or artifacts that have aesthetic attributes in some well-defined sense—most notably, literature, the fine arts, and building architecture. Technical or technological artifacts are seldom if ever described in stylistic terms—one does not conventionally associate style with the design of an oil refinery or a metallurgical plant. It is my intention in this chapter to point out that this convention need not hold and that at least in one technological domain—computer design—style has a central, essential place. Further, the development of a common science of design is likely to make obsolete this convention for many of the other design disciplines also.

The genesis of style is manifold. A most appropriate starting point for understanding the sources of style is Simon's (1975) assertion that it may originate in essentially three ways:

1. We identify styles simply by structural or behavioral features of a designed artifact. This is probably most evident in building architecture, where key structural features may be so heavily accented that even the layperson has little difficulty in identifying certain architectural styles—the Gothic, for example.

    Such distinctive features of the finished product may, in some contexts, be the hallmark of or have been inspired by a particular designer or school of design. It is because of this that the informed viewer may recognize a work of art by Rembrandt or Turner or El Greco.

2. We may also identify styles with the manufacturing process—that is, the method by which the artifact as conceived is shaped into reality. If we interpret the term "manufacturing" in a sufficiently wide sense to mean physical realization, then possibly no better example is provided of this component of style than computers, which have long been broadly categorized according to the technology used for implementing them. Since the technologies were also developed chronologically, there is a natural mapping of technological styles of computers to specific time periods.

    Thus, we now speak of "first-generation" computers as belonging roughly to the period 1946–1958, the "second generation" as spanning the period 1958–1964, the "third generation" systems as being developed between 1964 and 1972, and the "fourth generation" spanning 1972–1978, while a "fifth generation" is currently in its infancy.

    As is well known, each so-called generation is characterized primarily by a dominant logic technology—these being, respectively, the vacuum tube, the transistor, small- and medium-scale integrated circuits (up to about 100 gates per chip), large-scale integration (up to about 10,000 gates per chip), and very large scale integration (up to 100,000 gates per chip).

Notice that while the third generation technology rendered the second obsolete, just as the second made the first obsolete, at this time of writing, at least, the third, fourth, and fifth generation technologies actually coexist and, in a very real sense, offer a set of alternative implementation styles that can be chosen by the maker of computers.

To take another, quite different example, students of ancient metallurgy and technology, in their studies of the origin, evolution, and adaptation of metallic artifacts, frequently delineate stylistic differences between artifacts that are based on the different techniques (e.g., the lost-wax casting process, or the technique called "repousse") used by craftspeople for their manufacture.

3. Finally, it may be possible to associate style with the design process itself. In painting, the style of art known as pointillism is such a case. Its practitioners, notably Georges Seurat, mastered a technique, a method of using the brush, that gave rise to a recognizably distinctive design style.[1]

In the more familiar realm of software, the THE (Dijkstra [1968]) and VENUS (Liskov [1972]) operating systems may be said to be the products of a common style of design, in that both were created as a hierarchy of abstract machines, designed from the bottom up.

Since the concern of this book is with computer architecture as a design discipline, it is of considerable interest to inquire into the issues of style in computer architecture. In particular, we shall see that Simon's (1975) tridimensional categorization of style provides an unexpectedly appropriate framework for describing the architectural design process.

12.2 STYLE INDUCED BY ARCHITECTURAL FEATURES

Consider the first component of style—that which we associate with the features exhibited by an architecture. I shall call this simply *architectural style*.

Clearly, we can associate style with some very general characteristics of an architecture as, for example, when we distinguish between the "von

---

[1] Of course, in the terminology of the fine arts one does not conventionally refer to the act of creating a work of art as *design*. Within the framework of the present book, however, it seems perfectly reasonable to do so. Also it may be a matter of debate whether a technique such as pointillism should be regarded as a design or a manufacturing technique. The distinction, in the particular case of painting, is admittedly fuzzy.

Neumann" and "functional" styles. Thus, almost all general purpose computers that have been or are currently being constructed are of the former type, possessing essentially the following general features:

1. The use of an explicit instruction counter to define the locus of control through a computation.
2. A centralized control mechanism that realizes a well-defined, continuously looping cycle of operations in which instructions and data are fetched from main store into a centralized processing unit in order for instructions to be executed.
3. A system whereby change in the state of computation is effected by means of updating variables in main store (i.e., through side effects).

In contrast, in a functional style—represented most notably at the present time by data flow computers—there is no explicit locus of control as defined by an instruction counter. An instruction executes when and only when all its required operands are available. Further, the result of executing an instruction is a value that becomes an operand for one or more other instructions—that is, there are no side effects.

As in any such discussion, one may disagree about the precise group of features that characterize a style. For example, some authorities may impose the serial nature of instruction execution as an additional aspect of the von Neumann style. The set of features I have enunciated, on the other hand, means that the von Neumann style encompasses both serial and certain classes of parallel computers.

This brings to attention the well-known classification by Flynn (1966) of von Neumann machines into SISD, SIMD, MISD, and MIMD types. Leaving aside the MISD (multiple instruction stream, multiple data stream) category—which is more a theoretical possibility than a group of actual designs—it is clear that we have here a description of alternate endoarchitectural styles. For, one may distinguish between machines of the remaining three categories simply by pointing out the differences in their respective instruction cycles:

1. In an SISD (single instruction stream, single data stream) machine there is a single instruction cycle; operands are fetched in serial fashion into a single processing unit before execution of the instruction.
2. An SIMD (single instruction stream, multiple data stream) machine also has a single instruction cycle. However, many distinct sets of operands may be fetched to distinct processing units and be operated upon simultaneously within the single instruction cycle.
3. In an MIMD (multiple instruction stream, multiple data stream) machine several instruction cycles may be active at any given time, each independently fetching instructions and operands into distinct processing elements and operating on them in a concurrent fashion.

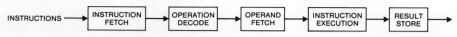

Figure 12.1 Instruction pipeline.

Of course, as in all matters of style, members of any one of these classes may differ considerably in other dimensions of the design space. Thus, for instance, the ILLIAC IV and the Burroughs Scientific Processor (BSP), both SIMD machines, differ at the exoarchitectural level: the ILLIAC IV is a general purpose SIMD machine while the BSP has a high-level language architecture oriented toward Fortran. At the endoarchitectural level, the arithmetic units in the BSP are pipelined, while those of the ILLIAC IV are not.

Pipelining is itself a distinct and extremely important architectural style that comprises a different dimension of the design space. The fundamental characteristic of a pipeline processor is that it is decomposed into distinct, dedicated subprocessors or stages that may concurrently execute different stages of different processes. Any particular process, however, goes through all the stages of the pipeline in strictly sequential fashion.

Pipelining may exist at different levels. For example, in the case of an instruction pipeline, the instruction cycle is decomposed into distinct stages, each performing a part of the instruction cycle of a distinct instruction. A given instruction would, of course, go through all the stages in serial fashion (Fig. 12.1). Thus at any time, given an n-stage pipeline, n instructions could be executing at the same time (Fig. 12.2).

It is possible, of course, for computers within both the SISD and SIMD classes to fall within the pipelining category; however, the notion of a style really emerges when the set of characteristics comprising the style dominates the architecture and influences its design. It is for this reason that, though the BSP employs pipelining in its arithmetic elements, it may not be considered an archetype of this style, since pipelining is not the dominating theme in its design. In contrast such machines as the CRAY-1, the Texas Instruments ASC, the CDC STAR-100, and the MU5 could be, and are, legitimately considered archetypal instances of this style.

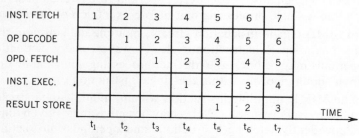

Figure 12.2 Pipeline timing diagram.

STYLE INDUCED BY ARCHITECTURAL FEATURES

Character data type
Move Compare

Figure 12.3 Character object.

As further examples of architectural style we consider some aspects of exoarchitecture. Here again we must think of these as an entirely different dimension (or a set of dimensions) of the design space, so that a particular style of exoarchitecture may well coexist with two quite different endoarchitectural styles in any given pair of computers.

I shall single out, firstly, two clearly delineated exoarchitectural styles as examples: the *stack-oriented* architecture exemplified by the Burroughs B6700/B7700 series and the *object-oriented* style represented by the Intel iAPX 432 processor.

In the B6700/7700 series, the use of the stack for the purpose of evaluating arithmetic expressions, controlling the addressing environment of block-structured languages, and managing storage during entry to and exit from procedures and blocks dominates the entire architectural design.

In the Intel 432, all entities defined by the hardware and usable by the programmer are structured into one or another type of objects—embodiments of the software notion of abstract data types. An object encapsulates in a single logical unit, a data type together with a set of operations such that given an instance of the type, the only way of manipulating the data item is through the defined operations. Thus as a simple example, one could view characters as objects consisting of the character data type together with a set of operations such as MOVE and COMPARE defined on them (Fig. 12.3). A slightly more complex example from the Intel 432 is that of a process object (Fig. 12.4), comprising a data structure that describes the state of a schedulable piece of software, along with an associated set of operations (e.g., READ PROCESS CLOCK, ENTER GLOBAL ACCESS SEGMENT).

Again, the notion of objects is the central and dominating theme in the 432 architecture, establishing a very definite style.

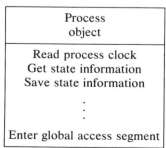

Figure 12.4 Process object.

Finally, one may mention a newly emerging architectural style. This is the so-called reduced instruction set computer (RISC) primarily advocated by Patterson and his coworkers in Patterson and Ditzel (1980) and Patterson and Sequin (1981). Apart from RISC I, currently being developed by Patterson et al., other RISC-like processors also under development include MIPS by Hennessey et al. (1982a, 1982b) and M3L by Castan and Organick (1982). The IBM 801 also has some similarity with these machines (Radin [1982]). Following Patterson and Sequin (1981), the RISC style may be said to be characterized by the following features:

1. A small set of relatively primitive instructions of essentially the same size is available.
2. Each instruction is executed in one machine cycle.
3. The instructions provide support for high-level languages and their compilers.
4. Only load/store instructions access memory. The other instructions operate upon registers.

Thus the RISC approach appears to represent a new exoarchitectural style, which for various reasons—notably simplicity, performance characteristics, and relevance to VLSI implementation—is likely to become an important design choice for architects in the future.

The examples given above are all, quite clearly, major issues of style that normally determine and establish the overall characteristics of an architectural design. However, subsystems may themselves manifest discernible and well-defined characteristics, to the extent that they establish their own styles. For example, a microprogrammed control unit is generally characterized in terms of its microinstruction organization being "horizontal" or "vertical."[2]

12.3 THE INFLUENCE OF ARCHITECTURAL STYLE ON THE DESIGN PROCESS

It is widely agreed that the design problem is characterized by two key features: (1) it involves the synthesis of a target object (a design) from a set of more primitive assemblies or components; and (2) it involves a search through an enormously large and complex space of partial designs and subassemblies for the appropriate set of components, whose proper assemblage would comprise a synthesis of the desired object.

---

[2]Note, though, that where the overall design objective is a microarchitecture, the choice of the microinstruction organization becomes a major design issue, establishing as it does one of the major stylistic components of the entire microarchitecture.

This search process is itself recursive: The selection of a component may in turn involve searching through the same combinatorial space for subcomponents whose appropriate assemblage would produce the required component.

A further characteristic of the design problem, as Simon (1975) has pointed out, is that there usually exists no finite algorithm for going directly to the solution. Of course, this is a characteristic shared by design and many nondesign problems—for example proving theorems.

In actual practice the designer does not explore the larger part of the combinatorially rich design space. At each stage of the design a process of selection operates, based on externally given constraints—the latter causing certain possible choices that do not satisfy the constraints to be automatically eliminated, thereby sharply reducing the size of the space. Among the choices left, the process of adding each of them to the partial design obtained up to that point and evaluating the resulting assembly against other constraints would possibly help to further eliminate some of the choices. To use the terminology of artificial intelligence, if we think of the design process as an instance of the generate-and-test paradigm (Simon [1975]) then constraints imposed by the generator (which in this case selects alternative components to add to the partial design) cause a reduction in the effective search space, while constraints associated with the test function reduce the effective design search space still further.

The existence of well-defined architectural styles provides a powerful basis for reducing the search through the design space, especially in the initial stages. A given design problem would be posed essentially in terms of certain (global) objectives and constraints that the design as a whole must meet. The architect, instead of initiating the search through the entire space of designs, partial designs, and assemblies, would select from a smaller space of architectural styles candidates that satisfy these global constraints. Notice that such a selection of styles would, in general, form an AND-OR set, for example of the form $(S_1 \vee S_2) \wedge (S_3 \vee S_4)$ where S_1, S_2 represent alternative styles that satisfy one set of global objectives and constraints, and S_3, S_4 constitute another. Thus a conjunction of these alternative pairs of styles is required to cover the set of global objectives and constraints.

In fact, at any stage of the design process, provided there exist discernible styles for the given (sub)problem, the choice of styles provides a basis for pruning off other branches from the search tree. Consider, for example, the standard method of stepwise refinement, or modular decomposition: the task at hand at any stage of the decomposition is to identify the components of the problem or subproblem. As in any case of such modular decomposition, an important decision is what criteria to use as a basis for decomposition.

For refining an architectural design problem into subproblems, I suggest that one appropriate criterion for such decomposition is that the resulting components or subproblems be associated with well-defined and discernible styles.

In summary, I suggest that architectural styles, in addition to defining broad, thematic features of architectures, can participate in the design process in two ways:

1. By reducing the search through the design space.
2. By providing a basis or a framework for decomposing a design problem into subproblems.

Example 12.1

In designing an exoarchitecture, suppose the initial objective is to produce a system for supporting most of the commercially available general-purpose programming languages (this objective may, of course, be expressed in terms of more specific subobjectives, such as an orthogonal and regular instruction set, support for procedure invocation and exit, support for structured data types, support for efficient expression evaluation, etc.). Then a particular range of architectures is immediately suggested—those falling within the so-called high-level language architecture style.

Notice that, as with all stylistic categories, individual member architectures may differ considerably in many respects; features may be borrowed or adapted from other styles and designs. However, as an initial design decision driven by externally set objectives, the beginning design space is sharply reduced, since other exoarchitectural styles are (at least tentatively) eliminated—for example, the so-called 360 style, characterized by a medium-sized set of general-purpose, multifunctional registers, a large, irregular order code, no generic operations, support for scalar data types only, and so on.

Example 12.2

In using the design automation system developed at Carnegie-Mellon University (CMU-DA) by Director et al. (1981), the designer may explore alternate design styles during the development of the register-transfer level data path structure.

Possible design styles in the CMU-DA system include a distributed style, characterized mainly by duplicate hardware for arithmetic and logic, low frequency of register sharing, and many independent paths; a bus style, which is characterized by a high degree of register and logic sharing and a centralized bus data path; and a pipeline style, characterized by the use of resources to maintain a constant flow of data.

The use of design styles in the CMU system, it should be noted, appear at a much lower level (i.e., at a later stage) in the design process than envisaged in the previous example—that is, after a behavioral description of the ar-

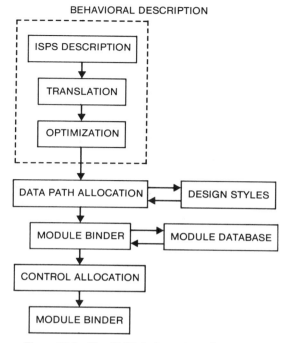

Figure 12.5 The CMU design automation system.

chitecture has been developed in ISPS, translated into an intermediate form and optimized. The main steps in the CMU design process are outlined in Fig. 12.5.

Example 12.3

The above example illustrates the point that in an explicit design process, stylistic choices may have to be made at certain stages when the preceding stages offer no guidelines as to what the next set of design decisions should be.

As another example of this situation, consider an outside-in design process in which we first develop the exoarchitecture and then the endoarchitecture. Here, we cannot apply the principle of stepwise refinement alone, since this principle taken literally means that each module or entity at the higher level is decomposed into a set of more primitive modules. Pure stepwise refinement, in other words, yields a strictly tree-structured hierarchy of modules in the sense of Fig. 1.1.

However, many facets of an endoarchitecture simply cannot be obtained as a refinement of exoarchitecture, since they exist independently of the latter (indeed, this is the basis for architectural families such as the System/

370 or the PDP-11 series). Obvious examples are the use of a cache memory, instruction pipelining, the presence of multiple ALUs, and so on.

Thus, the starting point of endoarchitecture design is only in part the overlying exoarchitecture. Much of the endoarchitecture (in particular, the data path structure and the precise nature of the control and data flow) must be developed by using independent criteria, and the possible space of initial choices may be enormous. Here again, the selection of a particular endoarchitectural style may help to reduce the effective design space.

12.4 THE INFLUENCE OF IMPLEMENTATION STYLE

We have already noted in Section 12.1 how one may categorize computers into generations according to the technology of implementation. The latter may also impose an influence on the architecture, so that architecture may itself be strongly affected by the various implementation styles at one's disposal.

Example 12.4

Recent work by Conway, Mead, and others (see Mead and Conway [1980]) has demonstrated that VLSI technology is most effectively used if the subsystems to be placed on a chip have an orderly, regular, repeatable structure, and the interconnection of the subsystems is also regular. Thus, computer endoarchitectures that can be implemented in terms of highly regular components will be favored over those that are less amenable to such implementation.

For example, Kung and his coworkers (Kung and Leiserson [1980], Kung and Foster [1980]) have developed a whole class of special-purpose structures called *systolic* architectures that are particularly suited for VLSI implementations. The basic idea underlying these architectures is to use a regular configuration of identical cells to implement a particular function or algorithm such that the entire complex can be placed on a single chip. It is envisaged that in future systems, a collection of such special-purpose processors would be connected through a central bus to a general-purpose system, as shown in Fig. 12.6.

Example 12.5

Consider the process of *emulation* of a target exoarchitecture on a microprogrammable host machine. If the target architecture is that of an existing machine, then clearly the emulation process—and the nature of the host microarchitecture—should be completely invisible as far as the user of the target architecture is concerned. This was the case, for instance, in Demco

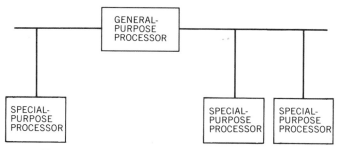

Figure 12.6 A system of special-purpose processors.

and Marsland's (1976) emulation of the PDP-11 exoarchitecture on the Nanodata QM-1.

On the other hand, one of the intended applications of dynamically microprogrammable computers is the implementation, either as prototypes or for production purposes, of new exoarchitectures oriented toward special applications or needs of the user. In such cases the architect may be required to take full advantage of the specific idiosyncrasies of the microprogrammable host in order to obtain an efficient implementation, to the extent that this influence may make its presence felt in the target architecture itself.

For example, Olafsson (1981) designed an exoarchitecture called the QM-C that is directly oriented toward the C programming language, but with the additional constraint that the architecture would be implemented by nanoprogramming the QM-1. The resulting exoarchitecture was influenced in several ways by this implementation style, most notably in the design of instruction formats, addressing modes, and data representations.

12.5 ARCHITECTURAL DESIGN STYLE

As noted in Chapter 2, architecture as a design discipline is at a rather primitive stage. Consequently, there has yet to emerge an established body of design methods or philosophies with which one may associate styles. This is in sharp contrast to the situation in software design, where such techniques as topdown, bottom-up or inside-out design, the data-oriented method developed by Jackson (1976), and the method of iterative multilevel modeling suggested by Zurcher and Randell (1968) are sufficiently distinct to lead to recognizably different design styles.

A few exceptions do exist, however. One general technique that has been applied by several architects with some variation in details is to base the design of an order code on a frequency analysis of operation and statement usages in high-level programming languages.

Instruction	p_i	Opcode	Opcode Length
A	0.50	1	1
B	0.30	01	2
C	0.10	001	3
D	0.03	00000	5
E	0.03	00001	5
F	0.02	00010	5
G	0.02	00011	5

Figure 12.7 Instruction frequencies and Huffman encodings (copyright © 1982, John Wiley and Sons; reprinted with permission).

One of the objects of such an exercise is to minimize the storage requirements for object codes; thus, by using the frequency of operations in a given sample of source code as an estimate of the probabilities of their occurrences, the number of opcode bits used to encode operations may be assigned in inverse proportion to the probabilities of the operations. Note that although an optimal usage of opcode bits may thus be obtained if opcodes are assigned by using Huffman encoding, the resulting variability of opcode lengths may be too large to make the optimal scheme practical. Less optimal—but satisficing—encodings may, however, be compared against the Huffman encoding to determine their relative "goodness."

We consider a specific example, taken from Myers (1982). Figure 12.7 shows a table giving the usage frequencies of a hypothetical set of instructions and its Huffman encoding. Note that if we were to select a fixed-size opcode field, its minimal length would be three bits. Hence a program of 1000 instructions will have 3000 opcode bits. Using Huffman encoding, the most frequent instruction A has a one-bit opcode, and the least frequently used instructions have a five-bit opcode. The average opcode length obtained according to the formula

$$\sum_{i=1}^{7} p_i L_i$$

where p_i, L_i are respectively the frequency and opcode length for instruction i—is 1.9 bits, so a 1000-instruction program would require only 1900 bits.

Clearly, for any realistic instruction set, the decoding complexity for the variably sized opcode field renders the optimal encoding impractical. A compromise between the fixed size and Huffman-encoded extremes may, however, be achieved, and one such compromise is given in Fig. 12.8, where it is assumed that opcodes can be either two or four bits in length. The average opcode length is 2.2 bits and the 1000-instruction program requires 2200 opcode bits.

Instruction	p_i	Opcode	Opcode Length
A	0.50	11	2
B	0.30	10	2
C	0.10	01	2
D	0.03	0011	4
E	0.03	0010	4
F	0.02	0001	4
G	0.02	0000	4

Figure 12.8 Compromise opcodes, after Myers (1982) (copyright © 1982, John Wiley and Sons; reprinted with permission).

Example 12.6

Both Myers (1982), in his design of the SWARD architecture, and Tanenbaum (1978) adopted the method described above for determining order codes. In particular, Tanenbaum proposed an experimental exoarchitectural design in which the order code, the instruction formats, and the instruction sizes were derived on the basis of a detailed, static analysis of a sample of experimental programs written in a typeless programming language called SAL. This analysis not only established frequencies for operators, but also for different types of assignment statements (e.g., statements of the form $X := X + 1$, $X :=$ constant, etc.) and higher-level constructs.

A similar approach was adopted by Olafsson (1981) in the work cited earlier on the QM-C architecture. Here, the basis for much of the order code and instruction format designs was a static analysis of several million lines of C code. The analysis determined usage frequencies for different operator and statement types, constant values, data types, and data structures. The main difference between Olafsson's work and the Myers/Tanenbaum approach was that in the former case, the frequency analysis was not used to determine opcode or instruction lengths, these being largely determined by the structure of the underlying host machine.

The design of the RISC I machine by Patterson and coworkers, described in Patterson and Sequin (1981), was also based on an initial static and dynamic analysis of source programs written mainly in C and Pascal. In this case, however, the main objective was to identify the frequently used operations that, when supported by the hardware, would lead to the greatest executional efficiency, rather than the most compact code.

This whole approach is an instance of a distinctive design style in which the architectural design is based directly on an analysis of the programming languages to be implemented on the machine. Indeed, this whole style already has a name—*language-directed architecture*. A closely related approach bases architectural design on an analysis of the classes of problems to be solved by the processor; we may term these *problem-directed architec-*

tures. Note that a special, degenerate, and—from the point of view of design methodology—not very interesting case is the design of special-purpose architectures dedicated to single applications.

Example 12.7

An instance of problem-directed design is provided by the initial stages of the MIMOLA design method, developed by Zimmermann (1980) and his colleagues, in which a specification of the target machine, expressed in the MIMOLA machine description language, is developed, starting with an initial set of sample problems from the target application classes. The general sequence of design steps is shown in Fig. 12.9.

Note that the estimated usage frequencies of the algorithms are introduced as what Zimmermann calls factors into the MIMOLA program set. Similarly, the final MIMOLA machine specification incorporates other initially established constraints and optimization goals as factors. This specification then becomes the input to a computer-aided design system (not indicated in Fig. 12.9).

Although architecture has not provided much in the way of design styles, some initial ideas have emerged. In particular, attempts have been made to adopt the standard techniques from the software domain to the realms of architecture and hardware. We shall consider here two examples.

Example 12.8

Recently, Leung (1981) outlined the structure of a 2×2 packet routing network designed according to a rigorous stepwise refinement process. As we saw earlier, this method involves the decomposition of a module into a set of more primitive, interconnected modules that provide a more detailed description of the entity being designed. This technique obviously lends itself very appropriately to the design of highly modularized, hierarchical structures.

Leung's approach applied to the routing network example can be summarized as follows: the starting point is a specification of the behavior of the hardware module in the architecture description language ADL, as described in Leung (1979). Next, the router is decomposed into four submodules and a specification of their interconnections. This is an instance of structural decomposition. The behavioral descriptions of the submodules are then developed. The result of structural decomposition is an ADL construct consisting of a description of the submodule interconnections, together with a behavioral description of each distinct submodule type.

Up to this point, the basic (atomic) data type was assumed to be the packet. In the next stage, the structure of packets as a sequence of bytes is developed; this is an example of data structure refinement. Since a data type

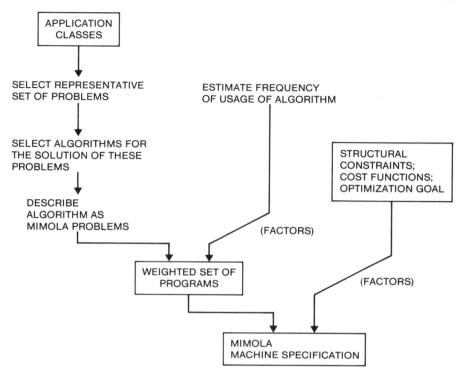

Figure 12.9 The MIMOLA design method, based on Zimmermann (1980).

is being replaced by more primitive types, the operations on the former must also be replaced by appropriate operations on the latter. Thus, as a consequence of data structure refinement, further behavioral refinement of the submodule types are needed. Since this also affects the interface between submodules, the original structural description of the router must also be updated. Note that that entire design is still represented in ADL.

Finally, the ADL description is transformed into an equivalent description involving only bit manipulations, after bit representations for the various data types have been selected. Leung refers to this as data representation refinement, although one may still view this as a further example of data structure decomposition. Again, this step results in further refinement of the behavioral description of the submodule types. This final ADL description is at a sufficiently detailed level for transformation into physical hardware.

Example 12.9

The QM-C machine (Olafsson [1981]) is a C-oriented exoarchitecture implemented on the QM-1 host. The design of this architecture involved several methodological issues that are discussed in various places in this monograph, particularly Chapters 13 and 14. It is especially characterized by its use of the two languages S*A and S*(QM-1) in the formal design process. Thus, the QM-C endoarchitecture is first described in its entirety in S*A. Then the process of stepwise behavioral refinement is applied to each individual S*A procedure to obtain the equivalent S*(QM-1) microprogram. The refinement process is guided and constrained by the available constructs in S*(QM-1). A more detailed discussion of this design process is presented in Chapter 14.

Finally, one should note that issues of design styles and methods are being very actively explored for lower levels of hardware design—especially in the realm of VLSI-based systems. The enormous complexity of state-of-the-art integrated circuits, and predictions of yet greater complexity of chips of the future, has prompted several research groups to investigate computer-aided design methods for chip design. Figure 12.4 shows one such approach, developed by Director et al. (1981) at Carnegie-Mellon University. Other techniques are being developed by Mead et al. at the California Institute of Technology (Mead and Conway [1980]) and at various other VLSI research centers.

12.6 BIBLIOGRAPHIC REMARKS

For this chapter, my primary intellectual debt is to the paper on design style by Simon (1975). For a discussion of the main stylistic movements in European (building) architecture see Pevsner (1963). Styles in the history of the art—particularly painting—are discussed in numerous sources; see for example Gombrich (1972) or the massive survey by Janson (1969). A superb symposium on ancient technology and metallurgy from the viewpoint of art is the volume edited by Doeringer et al. (1970).

Computer generations have been discussed in various sources. An early review is Rosen (1969), while for more recent reviews the reader may refer to Baer (1980) and Siewiorek et al. (1982). Baer (1980) also outlines the general principles of dataflow computers and discusses Flynn's (1966) well-known classification.

A recent review of the architecture of SIMD machines is given in Hwang et al. (1981). The single most authoritative and exhaustive treatment of pipelined computers (and likely to remain so for many years to come) is Kogge (1981). For shorter discussions, see Ramamoorthy and Li (1977) and Baer (1980).

A detailed description of the Burroughs B6700/B7700 series exoarchitecture is available in Doran (1979). An earlier, slightly different perspective on this innovative series is Organick (1973). While Intel (1981) is the official detailed specification of the iAPX 432 architecture, a brief discussion of its object-oriented features is given by Ziegler et al. (1981). Myers (1982) also discusses the iAPX 432 in some detail.

In spite of its recency, there is already a large body of literature on RISC architectures. The original ideas were presented in Patterson and Ditzel (1980). Subsequent papers include Patterson and Sequin (1981), Patterson and Piepho (1982), Fitzpatrick et al. (1981), Hennessey et al. (1982a, 1982b), Radin (1982), and Castan and Organick (1982). Patterson and Ditzel (1980) also included a critical assessment of complex instruction set computers (CISCs). For comments on this paper see also Clark and Strecker (1980).

For general reviews and perspectives on language-directed architectures, the reader is referred to Patterson and Ditzel (1980), Flynn (1980), and Myers (1982). Detailed discussions of specific architectures in this style include Tanenbaum (1978), Myers (1982), Patterson and Sequin (1981) and Olafsson (1981).

The already classic work on VLSI design is Mead and Conway (1980). The CMU-DA system has been described in various papers, including Parker et al. (1979), Thomas and Siewiorek (1981), and Hafer and Parker (1978). The most recent version of the system is discussed by Director et al. (1981). Various aspects of the MIMOLA system and its applications are available in Zimmermann (1980) and Marwadel (1981).

THIRTEEN

THE OUTER ENVIRONMENT

13.1 INTRODUCTION

From the viewpoint of a design discipline, exoarchitecture poses one of the more difficult problems in computer design. The main reason for this is that, of all aspects of computer design—whether at the architectural, register-transfer, or logic level—exoarchitecture is where the designer must come to terms with the needs imposed by the environment and simultaneously satisfy cost and technological constraints that are sometimes rather severe. In Simon's (1981) terms, exoarchitecture provides the interface between the computer's *inner environment*—its internal structure—and its *outer environment*.

Why should this pose a particularly difficult problem? It would not if the outer environment's range were within narrow limits and precisely predictable—as it may indeed be in the case of specialized applications. The problem arises in the case of computers that must operate in a wide-ranging environment or in an environment that cannot be predicted or characterized with any degree of precision.

The computer architect has traditionally resolved this problem by more or less ignoring it; or rather, by assuming and imposing a particular and sometimes artificial environment dictated primarily by the desired internal design of the computer, and formulating exoarchitectures that are adapted to this "virtual" environment. It then became the responsibility of compiler writers and operating system designers to bridge what Myers (1982) first called the

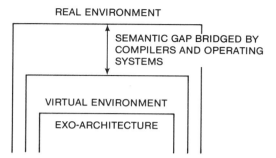

Figure 13.1 The semantic gap.

semantic gap between real and virtual environments (Fig. 13.1). The complexity of the design problem was transferred to the system programmers.

Myers (1982) is an original and full-scale treatment of exoarchitectures and the semantic gap problem, although a much earlier thinker along the same lines who has been largely ignored outside Britain until recent times is Iliffe (1972). To emphasize the new direction in exoarchitecture design, where the real rather than some artificial environment is taken as a factor, terms like *software-directed, language-directed,* and *high-level language* are used to describe such architectures. Needless to say, if designers had recognized the proper role of exoarchitectures to begin with, these prefixes would be redundant.

Thus, we see that the foremost issue in designing exoarchitectures is recognizing, characterizing, and predicting the shape of the outer environment. We shall return to this issue below. This is not, however, the only methodological problem.

13.2 ORDERING DECISIONS IN EXOARCHITECTURE DESIGN

The main components of exoarchitecture are the following:

1. *Storage organization:* the types and organizations of storage objects at the exoarchitectural level.
2. *Data types:* definition and characterization of the data types and their representation in storage.
3. *Addressing modes:* the different methods of specifying the addresses of stored objects.
4. *Instruction set:* specification of the structure and semantics of the instructions—that is, definition of the operational capabilities of the abstract machine that the exoarchitecture represents.
5. *Instruction formats:* representation of instructions in storage.

6. *Exception conditions:* specification of faults, traps, and interrupts and their consequences.

If we wish to obtain a formal design of an exoarchitecture then these components will ultimately be integrated into a single unified description in an appropriate language such as S*A or ISPS.

But before that stage there arises the question of what is the most appropriate way of partitioning the overall system into subsystems, and how should we order the decisions to be made in the course of design?

These are standard questions associated with the design and management of complex systems. They have, however, scarcely been addressed in the context of exoarchitectures. The problem is that the application of the principles of hierarchic design (which is one of the most powerful mental tools at one's disposal in managing complexity) presupposes that the system under development can be structured as a completely decomposable or at least nearly decomposable system (Simon [1981], Courtois [1977]).

The concept of nearly decomposable systems, as originally formulated by Simon and Ando (1961), is basically the idea that in many large and complex systems, variables can in some manner be aggregated into a small number of groups such that (1) the interactions among the variables within a group may be studied as if interactions among groups did not exist; and (2) interactions among groups may be studied without reference to intragroup interactions (Courtois [1977]).[1]

In the case of exoarchitectures we note that the partitioning of the architectural system into independent subsystems may not be applicable because of the strong interaction between the components.

Example 13.1

The primary factors influencing storage organizations are the different classes of data types and their representations, since these influence both the types of storage objects as well as the organization and size of storage words; and the instruction set and formats as these affect the organization and size of storage used as program memory. Conversely, the representation of data types is affected by the storage organization if the latter were to be determined (for some reason) before the data types.

For example, in the IBM System 370 (IBM [1981]) the storage organization is composed of a 32-bit word, byte-addressable main store and a set of registers, including 16 32-bit general purpose, 4 64-bit floating point, and 16 32-bit control registers. Clearly, the instruction set and format, on the one hand, and storage organization, on the other, interact intimately. For instance, a prior decision to include these two levels of storage determines in

---

[1] In the situation where the intergroup interactions are null the system may be said to be completely decomposable.

part the nature of the instruction formats and the opcode set. Furthermore, decisions as to the nature of the data types (e.g., integers, floating point reals, characters) to be supported by the architecture determine both the storage organization (e.g., inclusion of floating point registers) as well as the internal organization of main store words (e.g., viewing main store as an array of 32-bit words or as an array of bytes).

Example 13.2

There is an obvious, strong relationship between data types and the instruction set; the range and composition of the latter will depend on the composition of data types, and vice versa.

Example 13.3

Instruction formats will be directly affected by the instruction set, data types, and addressing modes.

Consider now the use of hierarchic design in constructing operating systems or other large-scale software, which is discussed in Dijkstra (1968) and Liskov (1972). The overall system is composed of a series of abstract machines $M_1, M_2, \ldots, M_n$ where the software running on machine M_i transforms M_i into M_{i+1}. In the classic case of the T.H.E. multiprogramming system (Dijkstra [1968]) the purpose of the abstract machine M_i is to create an abstraction or a virtual resource from some real hardware resource.

Courtois (1977) has discussed the criteria to be used in deciding on the ordering of such abstract machines. He identifies the following two design principles that should be obeyed in order to decide that the abstraction from a real resource R_i should be created at a lower level than the level of abstraction of another real resource R_j:

1. The abstraction from details of device R_i should be more convenient to program in order to create an abstraction from R_j than vice versa.
2. The use of an abstraction from R_i to program the abstraction from R_j should not prevent the latter resource from being controlled and allocated efficiently.

If the convenience of a tool or a concept is reflected in the frequency of its usage, then the consequence of the first principle above, Courtois concludes, is that the most heavily used abstract resources will be those implemented at the lower levels of the hierarchy. He also reasons that the consequence of principle 2 is that the fastest resources must be controlled and abstracted at the lower levels of the hierarchy.

We consider now the implications of these principles—in particular the first—for the design of an exoarchitecture where there exist strong or weak interactions between its constituents. I suggest that the principle of level-by-level hierarchic design as stated above may be applicable in this context. Note that we are not concerned with the construction of dynamic abstract machines but with the ordering of decisions concerning the architectural constituents.

For the exoarchitectural problem we adapt and extend Courtois' two conditions as follows.

Assuming interaction between components C_i and C_j, the design of C_i should precede the design of C_j (and thus be lower in the hierarchy) when any of the following conditions are obeyed:

Principle 1*

The design of C_i is determined by external factors that have higher priority over any internal conditions resulting from interactions between C_i and C_j.

Principle 2*

The component C_i is more convenient to use in the design of C_j than the converse.

Principle 3*

The design of component C_i lower in the hierarchy than C_j should not affect the efficient design of C_j.

Example 13.4

Consider the interaction between instruction format and storage organization. If it has already been decided that emulation on a given host will be the primary implementation style, then the word length of the host's program memory may dictate the design of instruction formats rather than vice versa. Thus the adoption of a particular implementation style may act as an external factor that overrides further considerations of the interaction between storage organization and instruction formats. That is, the instruction formats may be designed, given a particular memory word length (Principle 1*).

Example 13.5

Consider the relationship between data types and the instruction set. Clearly, it is more convenient to decide the specification and representation of data types first and then make decisions about the instruction set than vice versa, since it is the set of data types that determines the opcode composition and instruction semantics rather than the converse. Data types are the more independent components (Principle 2*).

Example 13.6

Suppose we have decided the nature of the storage organization before determining the instruction set or formats. Then the instruction length may turn out to be larger than the program memory word length. In that case we may rightly interpret the design of instruction formats as being inefficient if storage word bits are unused or several storage cycles are required to fetch a single instruction (Principle 3*).

In the next chapter, where we consider a full scale multilevel design problem, we shall attempt to apply these simple design rules in developing the exoarchitecture.

13.3 CHARACTERIZING THE OUTER ENVIRONMENT

In Section 13.1 it was pointed out that a primary issue in exoarchitecture design is determining or predicting the shape of the outer environment for which an exoarchitecture is being designed. Oddly enough, that this is an issue in architecture design came to be recognized quite recently in the evolutionary history of computers. Once the von Neumann machine model had stabilized, most exoarchitectural developments up to the early 1970s maintained its essential structure, augmented with features intended to reflect advances in technology, improve the raw speed of the machine, or provide capabilities that paid small regard to the convenience of their use or had very little to do with how such machines were actually used in computing environments.

13.3.1 Flynn's Study

An early study of the weakness of such well-known and influential exoarchitectures as the IBM 7090 and 360 systems was conducted by Flynn (1974). He began by partitioning the instructions into three broad classes: those that actually perform the data transforming operations (e.g., arithmetic, logical) observed in higher-level languages—these he termed "function" or *F-type* instructions; those that merely rearrange or set up data in preparation for other operations to be performed (e.g., LOAD or STORE), termed "memory" or *M-type* instructions; and "procedural" or *P-type* instructions involved in the transfer of control within programs (e.g., BRANCH, COMPARE, etc.).

In general M-type instructions in conventional third generation machines represent the overhead induced by the register-/accumulator-oriented characteristic of such architectures. Accordingly, as measures of code efficiency, Flynn proposed the following ratios:

$$\text{M ratio} = \frac{\text{number of M-type instructions}}{\text{number of F-type instructions}}$$

$$\text{P ratio} = \frac{\text{number of P-type instructions}}{\text{number of F-type instructions}}$$

$$\text{NF ratio} = \frac{\text{number of M-type + P-type instructions}}{\text{number of F-type instructions}}$$

$$= \text{P ratio} + \text{M ratio}$$

The last is the ratio of *nonfunctional* to functional instructions.

Clearly, these measures are to be treated with caution. As Flynn points out, whether a LOAD, a MOVE, or a COMPARE instruction is nonfunctional or not depends on the nature of the computation. For example, in a sorting program a COMPARE instruction must be treated as a prime functional operation while in a machine with index registers an ADD instruction may merely be acting as an overhead instruction in incrementing index registers.[2] Nevertheless, for certain job mixes these ratios do provide a quantitative idea of the nonfunctional overheads imposed by the environment on a certain class of architectures.

These ratios have been computed from instruction usage frequencies for several computers, notably the IBM 7090, IBM/360, DEC-10, and the PDP-11 in Flynn (1974, 1980); the corresponding NF ratios obtained were 2.8, 5.5, 2.6, and 6.3 respectively.[3]

Flynn (1974) has suggested several causes for the high nonfunctional overheads in the 360 and 7090 series, especially the former. Keeping in mind that the NF ratio is simply the sum of P and M ratios, a high M ratio clearly indicates high memory-register traffic, and therefore poor register usage.[4]

For the IBM 7090, the M ratio based on the Gibson mix was computed as 1.96. This overhead was attributed to the fact that the 7090 was a single accumulator machine, and though there were several index registers, there were no arithmetic facilities associated with the latter. Hence considerable data traffic was routed through the accumulator.

---

[2] In fact in Flynn (1974) it is pointed out that the above categorization was devised for technical (scientific) code only.

[3] It is not clear from Flynn (1980) what distribution of code yielded the ratios for the PDP-11 and DEC-10. However, the 360 and 7090 figures were earlier presented in Flynn (1974). The 7090 NF ratio was based on the "Gibson mix," which is representative of both technical compiler and object code, while the 360 ratio was obtained for a "general technical" mix consisting of 50% technical compiler, 50% object code. Thus the rest of this discussion is restricted to the 7090 and 360 analyses.

[4] It is to be recalled that registers serve as the fastest component of a three-level memory hierarchy—the other components being main and secondary memory—and the underlying assumption regarding their usage is that, once data is loaded into registers, a high proportion of computation is to apply to the contents of registers thereby avoiding as far as possible access to (slower) main memory and, as a consequence, reducing memory-register traffic.

The M ratio for the 360 machine was 2.9. Compared to the 7090, the 360 had considerably more general purpose registers, 16 in total. Thus, one would, a priori, expect much lower memory-register traffic. Yet there is a sharp deterioration in the M ratio. Flynn offers several reasons for this:

1. A significant change in the outer environment for the 360, in particular the tendency for late binding of variables to physical locations in main memory by techniques such as dynamic storage allocation and dope vectors. The former requires chaining through storage bits to ascertain space availability, while the use of dope vectors requires indirect addressing through multiple loading of registers.
2. The base register in the 360 is really an address extension register and necessitates additional loads and stores.
3. In the 360, some of the general purpose registers were reserved for specific purposes—either operating system usage or subroutine linkage. This effectively reduced the number of truly general purpose registers.

The P ratio components for the 7090 and the 360 series are respectively 0.81 and 2.5. Again, one notices the considerably higher procedural overhead in the case of the 360. Flynn points out that the 7090 COMPARE instruction performs both a test and a three-way branch, whereas the 360 COMPARE merely sets a condition code. Another reason for the high 360 P ratio is that its conditional branches are biconditional, whereas the actual path to be tested is often multidirectional. This results in chains of BRANCH ON CONDITION instructions.

In conclusion, insofar as the P, M, and NF ratios measure some aspect of architectural performance, the 7090 appears to have been better matched to its outer environment than the 360.

13.3.2 Analysis of Source Programs

Flynn's studies attempted to evaluate implemented computer architectures against their operating environments. A more direct way of characterizing the outer environment is by analyzing source programs written in high-level programming languages and measuring various source code characteristics that reveal precisely how languages are used.

Such empirical studies benefit the language designer, the compiler writer, and the computer architect. From the architect's point of view some of the important exoarchitectural questions answered by source code analysis are the following:

1. Identification of the opcode repertoire.
2. Relative importance of different opcodes (or opcode classes).

3. Determination of primitive and structured data types.
4. Composition and organization of storage.
5. Identification of addressing structure.
6. Design of instruction formats.

Indeed, source program analysis touches upon, and may influence the design of, almost every aspect of exoarchitecture.

The literature on the empirical study and measurement of programs is profuse, and we shall find it necessary to refer to one such analysis in the next chapter. The focus of our interest in this section, however, is to see to what extent one may abstract from these studies general properties of programs—properties that could provide an empirical foundation for architecture design.

Empirical data on source programs can be used in architecture design in two ways. First, if a particular language-directed architecture is being designed, then data pertaining to that language should obviously be available. This was the approach adopted, for example, by Tanenbaum (1978) in the design of his EM-1 architecture; by Flynn (1979, 1980) and his colleagues in Flynn and Hoevel (1983), Hoevel and Flynn (1977), and Neuhauser (1980) on their work on DEL (Directly Executable Language) machines; by Johnson (1979) in the development of a C-oriented 32-bit architecture; and by Olafsson (1981) in the design of the C-oriented QM-C machine. We shall see an application of Olafsson's data in the next chapter.

In contrast, if we are concerned with the design of a general language-directed architecture—one that is to be responsive to a broad range of programming languages and applications—of more interest are general (universal) characteristics of source programs. It is this issue that we address in this section.

The various qualitative and quantitative studies of source programs over the past years form the basis for some general hypotheses concerning the desirable properties of exoarchitectures. We describe these hypotheses below since, subject to further experimental corroboration, they can provide the necessary framework for the future design of architectures.

The hypotheses are based on data obtained from Alexander and Wortman (1975), Tanenbaum (1978), Myers (1977, 1978a, 1978b), Keedy (1978a, 1978b, 1979), Elshoff (1976), and Patterson and Sequin (1981). It is important, therefore, to take note of the general nature of the experiments conducted by these various investigators.

Tanenbaum's Experiment

A typeless, GOTO-less system programming language called SAL implemented on a PDP-11/45 is the source language. The executable statements are very similar to those of PASCAL and comprises the assignment statement and procedure CALL and RETURN, IF..THEN..ELSE..FI, WHILE..DO..OD, FOR, REPEAT..UNTIL..LITNU, DO FOREVER.. OD, EXITLOOP, CASE, and PRINT.

The basic data types in SAL are machine words (including general and input/output device registers), one-dimensional arrays of words and characters, bit fields, and programmer-defined data structures. There are two scope levels, local (stack storage) and global (static storage). A program consists of one or more procedures and zero or more modules that declare and initialize external variables.

The source programs, all written by the faculty and graduate students of the Computer Science Group at Vrije Universiteit in Amsterdam, were purposely designed as clean, well-structured, modularized programs. A specially instrumented compiler was used to collect measurements on over 300 procedures (with a mean of 18.2 executable statements per procedure) used in various system programs (the actual composition of the program mix was not provided). Static measurements were obtained by having the compiler count the number of occurrences of items in the source text. Dynamic measurements were obtained by having the compiler insert code into the object program to increment counters during program execution.

The Alexander-Wortman Experiment

The source language used was XPL, a dialect of PL/1 (McKeeman et al. [1970]), implemented for the IBM system/360. The composition of XPL statements include assignments, DO..END, DO VAR = E1 to E2..END, DO WHILE..END, DO CASE..END, CALL, IF..THEN, IF..THEN..ELSE, DECLARE, RETURN, procedure definition, and GOTO. Nineteen XPL programs were studied, including student compilers, as well as XCOM and ANALYZER, the two principal components of the XPL system.

Static measurements of source code characteristics were obtained by modifying the XPL compiler. Dynamic measurements were all related to System/360 instruction usage and obtained by interpretive execution of the object code. Dynamic characteristics of the source were not measured.

Olafsson's Experiment

The sample of source programs were all written in the C programming language described in Kernighan and Ritchie (1978) and obtained from the UNIX installation in the Computing Science Department at the University of Alberta in Canada. It consisted of the UNIX operating system, various compilers, loaders, assemblers, text-formatting packages, editors, utility programs, and several user programs. The total sample size exceeded 2.5 million lines of source code contained in 8905 procedures.

The data were collected by instrumenting the Portable C Compiler as described in Johnson (1978). Basically, this is a two-pass compiler that first performs lexical analysis, parsing, and symbol table construction, and builds parse trees for expressions. Most of the code generation is performed on the second pass.

The compiler was instrumented to collect measurements on a per-procedure basis; basically, only the machine-independent part of the compiler was instrumented so that the statistics reflected the characteristics of C usage independent of the target architecture.[5] Only static analysis was performed.

The Berkeley Experiment

As part of their design studies for the RISC 1 microcomputer, Patterson and his colleagues at the University of California, Berkeley conducted extensive experiments to determine the usage characteristics of C and Pascal as reported in Patterson and Sequin (1981). A set of four C programs and four Pascal programs were selected. The former included the Portable C compiler for the VAX, a VLSI mask layout program, a text formatter, and a sorting program. The latter comprised a Pascal P-code type compiler, part of a design automation system, a Pascal prettyprinter, and a file manipulation program. Statistics were gathered on the dynamic frequency of operands (integer constants, scalars, and array/structure types) and the dynamic usage of high-level language statements.

What are the important general conclusions we can draw from these studies concerning the shape of the outer environment for general purpose exoarchitectures? We answer this by proposing a number of qualitative propositions, providing empirical supporting evidence and, where applicable, statements of their implications for architecture design.

Proposition 1

The assignment statement is consistently one of the two most frequently occurring statements in high-level language programs, both in respect to static and dynamic frequencies.

Evidence

Table 13.1 summarizes data for the assignment statement from some of the published experiments. Except for the Patterson-Sequin (1981) figures for the C sample, no other statement type was found to have a higher frequency than the assignment.

---

[5] Of course, it must be noted that "C was originally designed for and implemented on the UNIX operating system on the PDP-11" (Kernighan and Ritchie [1978], p. ix). Hence to some extent, the language design has been biased by the PDP-11 architecture; it has, however, been implemented on various other architectures.

Table 13.1 Frequencies of Assignment Statements

Source	Language	Static Frequency	Dynamic Frequency
Tanenbaum (1978)	SAL	46.5	41.9
Alexander and Wortman (1975)	XPL	42	—
Olafsson (1981)	C	40.1	—
Patterson and Sequin (1981)	Pascal	—	36 ± 5
Patterson and Sequin (1981)	C	—	38 ± 5
Elshoff (1976)	PL/1	41.2	—

Proposition 2

The most common forms of the assignment statement are $A := B$, $A := k$, $A := A \text{ op } B$, and $A := A \text{ op } k$, where A, B denote variables or array elements and k is a constant.

Evidence

Tanenbaum's (1978) breakdown of the assignment statement by form is summarized in Table 13.2. There is, unfortunately, no further breakdown of the form with two right-hand-side (rhs) elements. Nor is there further explanation of "other forms with 1 rhs term."

Olafsson's (1981) analysis is shown in Table 13.3. (Recall that in C, the simple assignment statement "lvalue = expr" causes the value of "expr" to replace that of the object referred to by "lvalue," while an expression of the form $e_1 \text{ op } = e_2$ is equivalent to $e_1 = e_1 \text{ op } (e_2)$.)

Table 13.2 Breakdown by Assignment Type; Based on Tanenbaum (1978)

Assignment Statement Type	Static Frequency	Dynamic Frequency
var := constant	21.7	19.2
var := var	9.5	9.1
var := array element	4.3	3.3
array element := constant	4.1	2.8
array element := var	4.1	2.9
array element := array element	0.9	1.8
other forms with 1 rhs term	30.5	25.2
Total 1	75.1	64.3
Other forms with 2 rhs terms	15.2	20.4
Total 2	90.3	84.7

Table 13.3 Breakdown by Assignment type; Based on Olafsson (1981)

Assignment Statement Type	Static Frequency
A = B or A op = B	23.8
A = k or A op = k	35.5
A = B op k or A op = B op k	31.2
Total	90.5

Elshoff's (1976) data on the breakdown of the assignment statement indicate that 77.6% of all right-hand-side expressions had no operators (i.e., the statements were of the form A := B or A := k) and 20.5% had one operator (i.e., the statements were of the form A := A op B, A := A op k, A := B op C, A := B op k). The precise composition of the one-operator assignments was not given.

Implications

The architectural implications of these two propositions, particularly the second, are considerable. Note first that they provide a partial explanation for the high M ratios observed by Flynn (1974) for the IBM 7090 and 360 machines. More generally, proposition 2 brings into serious question the merits of both register-oriented and (arithmetic) stack-based architectures. Since the advent of the System/360 series the former has become the basis for a broad range of mainframes (e.g., the System/370), minicomputers (e.g., the PDP-11 series), and microcomputers (e.g. the Intel 8086), while the apparent utility of stacks for evaluating expressions has been a major reason for the implementation of hardware-assisted stacks in such major computers as the Burroughs 6700/7700 series (Doran [1979]).

In a lively exchange of notes conducted over a period of three years on the pages of *Computer Architecture News,* Myers (1977, 1978a, 1978b) and Keedy (1978a, 1978b, 1979) have debated the comparative advantages and disadvantages of pure stack-based, register-oriented, storage-to-storage, and accumulator/stack-based instruction sets.[6] Both Myers and Keedy con-

---

[6] The accumulator/stack-based architecture uses a single accumulator as an implicit operand for all arithmetic and logical operations. The second operand is either supplied explicitly with the instruction—in the form of a constant or a memory address—or related implicitly to the stack top. Thus, a statement such as A := A + B would generate code of the Form LOAD A; ADD B and only the accumulator would be required. The stack would only be used when intermediate results have to be stored, for example in executing the (much less frequently occurring) A := (B+C) + (D−E). This technique is used in both the MU5 (Morris and Ibbet [1979]) and the ICL 2900 series (Buckle [1978]).

clude that for the very common types of assignment statements, both register- and pure stack-oriented architectures consistently perform poorly compared to the other two. The performance measures used were:

1. The number of instructions.
2. The number of storage bits.
3. The number of distinct instruction elements (e.g., opcodes, addresses) that must be decoded by the processor.

Table 13.4 summarizes the Myers-Keedy data for three of the common forms of the assignment statement, while Table 13.5 presents two additional statements, the first representing the exchange operation and the other a larger but less frequently occurring expression.

Remarks

1. The measures are obtained by assuming that opcodes are eight bits in length, register addresses four bits, and storage addresses 20 bits.
2. The PUSH & LOAD operator pushes the accumulator value onto the stack top and loads the accumulator with the memory operand. Conversely, the STORE & POP operator stores the accumulator value in the memory address and reloads it from the stack top which is popped. The normal arithmetic operators ADD, MULT, and so on perform arithmetic operations on the memory operand and the accumulator value and leave the value in the accumulator. However, the MULT(TOS) causes the values of the accumulator and top of stack to be multiplied; the result is left in the accumulator and the stack top is popped.
3. The figures in parentheses in the storage-storage column show the effect of adding an 8-bit *length* field to appropriate instructions.

In summary, these data indicate the general inferiority of general registers and expression stacks—two well-entrenched textbook notions—as exoarchitectural features. On the evidence provided, the storage-to-storage technique (advocated by Myers) appears marginally superior to the accumulator/stack technique (supported by Keedy) for the smaller, more common statements, while the converse holds for the exchange and larger assignments.[7]

---

[7] It is important to note that the above discussion only applies to the use of stacks for expression evaluation. The use of control stacks for subroutine management is not under consideration or debate here. As Myers rightly points out, control stacks can be used with any type of instruction set.

Table 13.4 Summary of the Myers-Keedy Data on Common Assignment Types

Source Statement	Stack		Object Code			Storage-Storage	Accumulator/Stack	
			Register					
(1) A := B	PUSHAD	A	LOAD	R1,B		MOVE A,B	PUSH&LOAD	B
	PUSH	B	STORE	R1,A			STORE&POP	A
	STORE							
Instructions	3		2			1	2	
Size	64		64			48 (56)	56	
Elements	5		6			3 (4)	4	
(2) A := A+B	PUSHAD	A	LOAD	R1,A		ADD A,B	PUSH&LOAD	B
	PUSH	A	ADD	R1,B			ADD	B
	PUSH	B	STORE	R1,A			STORE&POP	A
	ADD							
	STORE							
Instructions	5		3			1	3	
Size	100		96			48 (56)	84	
Elements	8		9			3 (4)	6	
(3) A := B+C	PUSHAD	A	LOAD	R1,B		MOVE A,B	PUSH&LOAD	B
	PUSH	B	ADD	R1,C		ADD A,C	ADD	C
	PUSH	C	STORE	R1,A			STORE&POP	A
	ADD							
	STORE							
Instructions	5		3			2	3	
Size	100		96			96 (112)	84	
Elements	8		9			6 (8)	6	

Table 13.5 Myers-Keedy Data on other Types of Assignments

Source Statement	Stack	Object Code — Register	Storage-Storage	Accumulator/Stack

(4) A := : B	PUSHAD A
	PUSH B
	PUSHAD B
	PUSH A
	STORE
	STORE
		LOAD R1,B
		LOAD R2,A
		STORE R1,A
		STORE R2,B
			MOVE TEMP,A
			MOVE A,B
			MOVE TEMP,B
				PUSH&LOAD A
				PUSH&LOAD B
				STORE&POP A
				STORE&POP B

Instructions	6	4	3	4
Size	128	128	144 (168)	112
Elements	10	12	9 (12)	8

(5) A := (B+C)*
(D−E)
	PUSHAD A
	PUSH B
	PUSH C
	ADD
	PUSH D
	PUSH E
	SUB
	MULT
	STORE
		LOAD R1,B
		ADD R1,C
		LOAD R2,D
		SUB R2,E
		MULT R1,R2
		STORE R1,A
			MOVE A,B
			ADD A,C
			MOVE TEMP,D
			SUB TEMP,E
			MULT A,TEMP
				PUSH&LOAD B
				ADD C
				PUSH&LOAD D
				SUB E
				MULT (TOS)
				STORE&POP A

Instructions	9	6	5	6
Size	172	176	240 (280)	148
Elements	14	18	15 (20)	11

Proposition 3

The ADD (including INCREMENT) operator is usually—and by a considerable margin—the most frequently used arithmetic or logical operator, followed by SUBTRACT, AND, and OR. Among the relational operators, the most common—again by a considerable margin—is EQUAL, followed by NOT EQUAL.

Evidence

Olafsson's (1981) analysis of C programs indicates that the ADD operation accounts for 53.6% of all arithmetic operations, followed by 12.1% for SUBTRACT. Among relational (which in his analysis include logical) operators, EQUAL accounts for 35.2%, NOT EQUAL for 13.2%, and AND for 12.6%.

Ditzel's (1980) data on SPL, the Symbol Programming Language, shows 27% ADD statements and 11.7% SUBTRACT statements. Fifty percent of all relationals are EQUAL, while NOT EQUAL, GREATER THAN, and LESS THAN have roughly the same frequencies, 12–13%.

Elshoff's (1976) analysis of PL/1 programs shows that ADD accounts for 68% of all arithmetics followed by 16% for SUBTRACT. Of relational operators, 66% are EQUAL, followed by 18% NOT EQUAL. Tanenbaum's (1978) sample of SAL programs includes 50% ADD followed by 28.3% SUBTRACT among the arithmetics; among the relationals, EQUAL accounts for 48.3%, followed by 22.1% of NOT EQUAL.

Implications

A major objective of exoarchitectural design is to minimize memory space requirements for programs. Since operations do not have equal usage frequencies, one way of attaining this objective is to construct variable length opcode fields with the most frequent operations (e.g., ADD, INCRMT, EQUAL, NOT EQUAL) encoded in the smallest fields, rather than encoding all operations in the same fixed-length field.

This approach has been discussed previously by Tanenbaum (1978) and Myers (1982, Chapter 17). As Myers has noted, given the estimated probabilities of operations, the method called Huffman encoding will generate the optimal encoding of opcodes. However, in practice, Huffman encoding is never used since it may produce a very large number of opcoding lengths, thus greatly increasing the complexity of opcode decoding during instruction interpretation. Instead, a few fixed sizes for the opcode field are selected, with the most frequent operations assigned the smallest field. This technique was used, for example, in designing the Burroughs B1700 COBOL/RPG and

B1700/SDL exoarchitectures (Myers [1982], Chapters 12 and 17; Wilner [1972]).[8]

In the former case the seven most frequent instructions have three-bit opcodes and the remaining instructions have nine-bit opcodes. In the latter instance, the SDL-oriented exoarchitecture incorporates three classes of opcodes, of lengths four, six, and 10 bits. The most frequent operations are encoded in the smallest field and the least frequent ones in the largest.

Note that because the Huffman method produces an optimal encoding, the efficiency of a given (nonoptimal) encoding scheme relative to the corresponding Huffman encoding can be measured by comparing the weighted average number of bits for all operations resulting from the two schemes. Here, again, we see that the architect must be content with producing a satisficing rather than an optimal design.

Proposition 4

The single most important control statement is the procedure CALL. It is also the most time-consuming of the high-level language operations.

Evidence

The evidence for this proposition is not very consistent across all the studies. Moreover, the second part of the proposition is based on a single set of data and must undoubtedly be subject to further tests.

Olafsson's (1981) static analysis of C programs indicates that 26.4% of all statement types are the procedure CALL, the second most common statement type. Moreover, in 77.3% of all procedures, the ratio of the number of executable statements to CALL statements is less than 4.5; in only 17% of cases does this ratio exceed 6. This suggests a high level of modularization in Olafsson's sample.

Tanenbaum (1978) found that 24.6% of all statements were the procedure CALL; the dynamic frequency of the CALL was, however, much less: 12.4%.

A more telling analysis is presented by Patterson and Sequin (1981). Their dynamic frequencies for Pascal and C CALL statements were respectively 12 ± 1% and 12 ± 5% (which agrees with Tanenbaum's [1978] data). To determine which statements consumed the most time in executing typical programs, they examined code produced by typical versions of each statement based on data provided by W. A. Wulf from his studies of compilers. By multiplying the frequency of occurrence of each statement with the cor-

---

[8] SDL is the language used to write the operating system and compilers for the B1700. These programs are executed by the SDL-oriented exoarchitecture.

responding number of machine instructions and memory references, the weighted frequencies of CALL/RETURN statements were determined to be 43 ± 4% for Pascal and 45 ± 19% for C. Thus the CALL/RETURN combination was found to be the most time-consuming of all the high-level language operations.

In contrast to the above data, Alexander and Wortman (1975) found only 13% of XPL statement types to be CALL statements, while only 2% of the statements in PL/1 programs studied by Elshoff (1976) were CALL statements. Elshoff's sample had, however, a high (nearly 12%) frequency of GOTO statements, suggesting that the sample programs were relatively unstructured.

Implications

If we accept the data provided by Olafsson (1981) and Patterson and Sequin (1981), architects must pay particular attention to the efficient implementation of procedure CALL and RETURN operations, since the execution time efficiency of a computer is greatly affected by these operations. Conventional architectures normally require a sequence of machine instructions for effecting the context switch involved during procedure entry and exit: saving and restoring registers, passing of parameters and results, and the allocation and deallocation of storage for local variables. Some of the recent architectural proposals, for example, incorporate powerful CALL and RETURN instructions that, coupled with appropriate storage organizations, help reduce both program size and execution time; see Myers (1982), Olafsson (1981), Patterson and Sequin (1981), and Wirth (1981).

13.4 OTHER ASPECTS OF THE OUTER ENVIRONMENT

In the preceding section I have reviewed some of the published studies that, collectively, provide a partial map of a computer's outer environment. From these, a few propositions have been suggested that seem sufficiently supported by the evidence to serve as a rational basis for the design of some aspects of exoarchitectures.

The outer environment is, of course, far richer than what might be suggested from the foregoing discussion. In particular, there are several issues pertaining to current thinking in programming methodology, language design, and software reliability that lend further shape to the outer environment. These issues include:

1. The notion of data types and their architectural implementations in the form of self-defining or tagged data.
2. The principle of data abstractions—that is, packages or modules encapsulating a class of data objects together with a set of procedures or

operators that operate on them—and their implications for exoarchitectural design.
3. The implications of high-level concurrent programming concepts, for example, process communication and synchronization.
4. Storage organizations and addressing modes.
5. Storage protection.

Many of these (and several other) concepts, and their implications for exoarchitecture design, have been discussed in great detail by Iliffe (1972, 1982), Dennis et al. (1979) and, in particular, Myers (1982). Recently, Wulf (1981) has also considered the problem of exoarchitectural design from the compiler writer's point of view, while Denning (1978) has pondered the implications of operating system requirements for the computer architect. However, it seems reasonable to assert that much more systematic thinking must be invested in both qualitative and quantitative characterizations of these general aspects before we can extend the kind of propositions stated in Section 13.3.2 to these issues. From the point of view of computer architecture design, then, an open research problem is the identification and construction of a knowledge base of general propositions concerning the shape of the outer environment.

FOURTEEN

DESIGN OF AN ARCHITECTURE: A CASE STUDY

14.1 TOWARD A DISCIPLINE OF COMPUTER ARCHITECTURE

A conventional textbook on computer architecture attempts to present its author's view of the state of the knowledge in the field. In a sense, both the form and the content of such a text is greatly influenced by the tradition of textbook writing in the natural sciences; architecture is treated much in the same way that biologists view organisms or astronomers galaxies—as entities that are given and that need to be described, classified, analyzed, and explained.

However, computers are not naturally occurring phenomena; they are sophisticated and complex artifacts. Computer architecture deals with their *design* as well as their description and analysis. Its seems desirable, therefore, in order to reflect the true nature of its subject matter, that the future textbook eschew the natural science–minded view of architecture and treat it instead for what it actually is—a study of the artificial (Simon [1981]). Accordingly it is up to the theoretical and practicing computer architects to make continuous contributions to a discipline of computer architecture that will be concerned not only with architectures as given entities, but also with the process of designing such entities.

A *discipline,* according to the *Concise Oxford Dictionary* means, among other things, a "branch of instruction or learning" and also a "system of rules for conduct."[1] Clearly, both these interpretations are relevant to our context. A discipline of computer architecture is, then, a branch of learning dealing with the logical aspects of computer design; it also refers to a system of rules and principles governing the design of computer architectures.

At the present time one would hesitate to make claims to a *science* of computer architecture for the following reason: in current epistemology "science" as a term is ineluctably associated with the study of naturally occurring phenomena; as such, established historico-philosophical attributes of the "scientific method" such as Popper's (1965, 1968, 1972) criterion of refutability as a hallmark of scientific propositions or Kuhn's (1970) theory of scientific paradigms do not apply without substantial revisions to the domain of the artificial. (In fact, what is urgently needed is a philosophical examination of the logic and methodology of the artificial sciences.)

In the absence of such a philosophical root I shall forbear to use the term "science" because of the utter confusion existing in some people's minds about the scientific method and the role of creativity in science and in the arts. The word "discipline," insofar as it admits the notion of the rational into our discussion, seems more appropriate if only because it is more prudent.

In this book I have discussed several issues that I believe are important components of a prospective discipline of architecture. These may be listed as follows:

1. The concept of architectural levels.
2. Hierarchic organizations.
3. Formal description of architectures.
4. Design verification.
5. Style in architecture, design, and implementation.
6. Heuristic design.
7. Principles of computer architecture.

Note that much of mainstream research and development in architecture has so far been confined to the topic of architectural principles (item 7), though the importance of such issues as design representation and style has recently been recognized.

It will also be noted that while parts of this prospective discipline are formal, as when we are concerned with architecture description and design verification, other parts are distinctly soft, involved as they are with description and classification—as, for example, when we discuss architectural levels or style.

The ambition of every practitioner of a scientific discipline is to elevate it to the realm of the hard sciences. In many other fields this has resulted in a

---

[1] J. B. Sykes (Ed.), *Concise Oxford Dictionary*, 6th ed. Oxford: Clarendon Press, 1976.

general air of unreality, since formalization is frequently achieved through convenient approximation and elimination of intractable aspects. Thus I do not expect a future discipline of computer architecture to be entirely hard in the sense that physics or classical engineering is hard. Rather, it will more resemble the qualitative natural sciences of the older kind (e.g., geology, botany) or an artificial science such as building architecture: a composite of formal techniques and tools on the one hand, and informal *but rational* concepts on the other. The sine qua non for an architectural discipline, however, must be the elimination of the ad hoc at every stage of the design process, along with the construction of a strong formal component for the description and verification of designs.

The aim of this final chapter is to tie together the various strands developed in this book by describing the design process for a relatively large, nontrivial architecture. For this purpose, the selected example is based on the work done by Olafsson (1981) and reported briefly by Dasgupta and Olafsson (1982) on the multilevel design of a language-directed architecture.

14.2   TWO POINTS TO PONDER

In Chapter 3 it was pointed out that in order for computer architecture to emerge from the craft stage it should satisfy the characteristics of the formal design process. In fact, the whole purpose of Chapters 5 through 8 was to provide instances of formal design.

However, as we have just noted—and previously elaborated in Chapters 12 and 13—the design of architectures (or indeed any complex system intended to serve some purpose in the real world) involves factors other than those encapsulated in the notion of formal design. For example, it involves knowledge of architectural principles; of the different styles in design, implementation, and architectural subsystems; and of ways of organizing our thought processes.

The question arises of how, in the course of the design process, these formal and nonformal components can come together to produce a rational design. This is an enormously complex and poorly understood problem that touches on issues in cognitive psychology and artificial intelligence.

An exploration of this problem is certainly necessary if we wish to build computer-aided design systems for architectural synthesis, since we must establish the extent to which we could realistically automate the design process. At the same time, a proper understanding of the interaction of informal and formal design may help to identify the form and content of a design data base for such systems. For example, I suggested in Chapter 13 that an explicit recognition of the notion of style could provide a rational basis for decision making during the design process. Thus, a design data base could encapsulate facts or knowledge concerning architectural, implementation, and design styles and their interactions.

A second point worth considering is the following. That design is an act of synthesis is obvious. However, this is a gross simplification of the matter. For it has also been widely observed that design is, in fact, both analysis and synthesis: the act of analysis decomposes the problem into subproblems (and the latter into still more primitive subproblems, etc.); synthesis then takes over the process of assembly.

Some of the questions that arise in this context are what kind of criteria should be used in decomposing the target design into components? How should one continue the decomposition process before starting the task of assembly?

These questions have been widely addressed in the domain of software. For example, Parnas (1972) has suggested that for decomposing a program into modules—a design decision undertaken before work on the modules begins—a criterion that may be used is information hiding. That is, each module would be characterized by one or more critical decisions that are hidden from others.

The best-known response to the second question is known as the principle of stepwise refinement, originated, so far as I can tell, by Dijkstra (1972) and Wirth (1971, 1973). This notion so dominates our thinking on the program development process and has such wide implications for design in general that it is worth stating in full. A concise formulation is found in Alagic and Arbib (1978, p. 15):

> Decompose the overall problem into precisely specified subproblems, and prove that if each subproblem is solved correctly and these solutions are fitted together in a specified way then the original problem will be solved correctly. Repeat the process of "decompose and prove correctness of the decomposition" for the subproblems; and keep repeating this process until reaching subproblems so simple that their solution can be expressed in a few lines of a programming language.

The general idea, then, is to continue decomposition (analysis) until subproblems are obtained that can be solved by using predefined building blocks (in the case of programs these are constructs in the target programming language). The task of synthesis or assembly is then to compose and interconnect the building blocks so as to satisfy the desired function of the overall system. The UCLA design method discussed in Estrin (1978) is an example of the explicit application of this approach (Fig. 14.1).

The principle of stepwise refinement appears to have validity across a wide range of design domains. However, when we consider computer architecture it is not particularly clear what the predefined building blocks should be in this context. Nor is it obvious that the Parnas criterion can be applied mutatis mutandis to architecture designs. The design process discussed below is intended to shed some light on these issues.

216   DESIGN OF AN ARCHITECTURE: A CASE STUDY

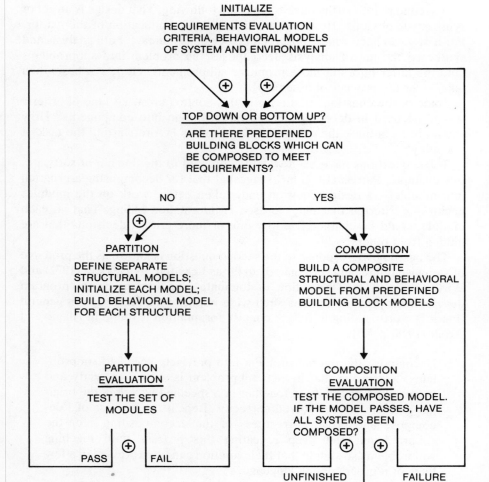

Figure 14.1 The UCLA design method (copyright © 1978, AFIPS; reprinted with permission).

14.3 DESIGN OF THE QM-C ARCHITECTURE

Our objective is to design an exoarchitecture that will directly support programs written in the C programming language (Kernighan and Ritchie [1978]) and be implemented through emulation on the Nanodata QM-1 (Nanodata [1979]). The resulting (abstract) machine is named the QM-C.

In pondering this design problem we first note that the problem itself implies a certain architectural style (a language-directed architecture) and a

DESIGN OF THE QM-C ARCHITECTURE

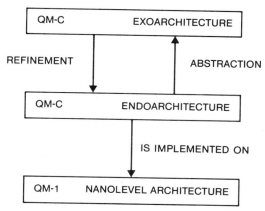

Figure 14.2 Architectural levels in the QM-C.

certain implementation style (emulation on a given microprogrammable host), thereby sharply reducing the space of possible designs to be searched. The architect need only consider architectures oriented toward C; among the resulting possibilities, only a subset fit for implementation on the QM-1 are candidates for further consideration.

Remark. Strictly speaking, the above paragraph is misleading: the space of possible designs does not exist a priori, and neither does the designer search through such a space. It is his or her task to generate an architecture that represents some point in the design space. The given objectives serve to eliminate right from the onset certain possible (perhaps partial) designs that the architect may otherwise have generated.

How can the designer proceed in a rational fashion from the given objectives to a final solution? As a first step we can structure the problem as a set of subproblems, using the principles of hierarchies discussed in Chapter 1. I noted there that in the domain of computer architecture hierarchies may exist in three ways: (1) where the hierarchic relationship between adjacent levels is one of "consists of" (Fig. 1.1); (2) where the relationship is one of "abstraction" (Fig. 1.2); or (3) where it is one of "is implemented on" (Fig. 1.3).

Applying these notions, the design problem may be organized according to the hierarchy shown in Fig. 14.2.

In the QM-1 there are, in fact, two levels of microprogramming (Fig. 11.1). At the upper level, *vertical microprograms* reside in control store and interpret instructions residing in main store. In QM-1 terminology, the architecture of the machine as seen by the control store programmer is called its *microarchitecture*. At the lower level, *horizontal nanoprograms* reside in nanostore and interpret (micro)instructions residing in control store. The corresponding architectural level is termed the *nanoarchitecture*. The hierarchy of Fig. 14.2 was constructed on the premise that object programs compiled from C would reside in control store (i.e., C object programs would really be QM-1 microprograms), and hence the base architecture for the emulator would be the nanoarchitecture. The decision to store C object programs in control (rather than main) store was dictated by the architect's (i.e., Olafsson's) desire for high-speed interpretation of C object-code.

The next step involves the choice of a design style that would direct the remaining phases of the design process. As noted in Chapter 12, the contemporary architect is at somewhat of a disadvantage in this regard, since he or she does not have an established catalog of styles to choose from. However, given that the general problems of design in architecture and software are rather similar, one can adopt the techniques known to the software designer.

The very nature of the problem at hand—namely, the design of a *language directed* architecture—suggests an obvious design style, the outside-in approach described by Freeman (1980). In the present context, given the hierarchic decomposition of the problem as shown in Fig. 14.2, this style is characterized by the ordering of design decisions proceeding from the outermost layer of the system (i.e., the exoarchitecture) through the intermediate layer (the endoarchitecture) to the implementation (the nanoprogram). Note with reference to Fig. 14.2 that since the QM-1 nanoarchitecture is given, it will only appear in the decision making process as a factor in the design of higher architectural levels.

Remark. It is appropriate to emphasize here the distinction between outside-in and topdown (or stepwise-refined) design styles. As noted in Section 14.2, in the topdown approach, design decisions proceed from the most abstract to the most concrete. Thus, this approach can be adopted in developing the detailed design of the individual architectural levels, while the overall design process follows the outside-in style.

14.3.1 Exoarchitecture Design

Choice of Implementation Style

In Chapter 12 I discussed the influence of style selection on reducing the effective space of designs to be searched. Clearly, at the onset of design, decisions that effectively prune the design space are preferable; such decisions will be the more influential ones and should be made first. (We may, of course pay a penalty for this; if at some later point this decision is changed the design process may have to start over.)

For the present problem, the most important decision was whether the target language or the host machine should be the primary determining influence on the QM-C architecture. A brief consideration of these two possibilities indicated that the latter had considerable advantages as far as ease of implementation and performance were concerned. Thus, it was decided

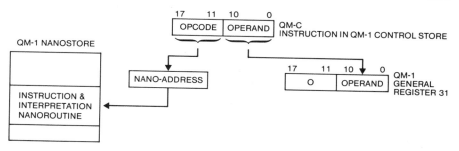

Figure 14.3 QM-C instruction decoding mechanism.

that a C-oriented exoarchitecture would be designed within the limits imposed by the QM-1's nanoarchitecture; the implementation style became the most influential decision. It fixed the word length of the QM-C at 18 bits, placed restrictions on the instruction formats (seven-bit opcodes, and restricted combinations of 5-, 6-, 11-, 12-, or 18-bit operands), and mapped the general structure of the instruction decoding mechanism naturally onto the available hardwired microinstruction decoding mechanism in the QM-1 (Fig. 14.3).

Mapping the Outer Environment

Even though the QM-C architecture is to be constrained by the host machine's (nano)architecture, this still leaves much room for the C language to influence the design. In order to determine the shape of the outer environment, Olafsson gathered extensive data on the actual use of the C language. This experiment, and some of the important statistics he obtained, have already been described in Chapter 13.

Decomposition of the Informal Design Problem

In decomposing the design problem into subproblems it must be realized that this may take place in quite different ways, depending on whether we are in the informal or formal stages of design. The latter stage will come later in this design process, so I shall defer the formal decomposition problem for the present.

An exoarchitecture, to paraphrase the definition stated in Chapter 1, is the total of a computer's features and capabilities required to be known to the outside world. For a von Neumann–style machine the primary components of exoarchitecture are the following:

1. Storage organization.
2. Data types.
3. Instruction formats.

4. Instruction set (order code).
5. Addressing modes.
6. Exception conditions (interrupts).

Thus at the informal stage of design, the problem is to make decisions concerning each of these components. In addition, for the particular implementation style selected—emulation—the nature of the mapping of storage organizations and data types onto the host machine's storage classes becomes an additional design decision.

As pointed out in the previous chapter, the above exoarchitectural components are by no means independent of one another. Indeed some of them are strongly related—note for example the dependence of instruction formats on storage classes, the instruction set, and addressing modes or the mutual interaction of data types and the instruction set. Thus some sort of ordering of the sequence of decisions must be imposed. One possibility is a decision hierarchy with the characteristics that:

1. Hierarchic levels are ordered from the bottom up.
2. Decisions on a component at level i of the hierarchy can be taken independently of decisions on other components at the same level.
3. Decisions on components at level j can only be influenced by decisions at levels $j - 1, j - 2, \ldots$.

Figure 14.4 shows the decision hierarchy for this particular design. Notice that decisions concerning the instruction format precede the design of the instruction set. This is because in this particular design problem the host architecture dictates to a large extent the general nature of efficient instruction formats. Thus it makes sense to establish these before deciding on the order code.

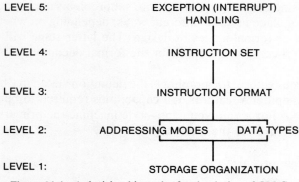

Figure 14.4 A decision hierarchy for the design of QM-C.

Informal Design of Exoarchitecture

The informal design of the QM-C exoarchitecture follows a fairly conventional procedure. However, several aspects of this design stage should be emphasized.

First, the so-called creative part of the design process, to the extent it is present, is essentially restricted to the informal design phase. Creativity appears to be manifested in two ways:

1. Where the design necessitates the invention of new architectural principles—that is, where the designer is also undertaking research. This form of creative or innovative design is encountered in the literature whenever novel, experimental, or paradigmatic architectures are being developed. See for example, the Cambridge CAP computer in Wilkes and Needham (1980), the ILLIAC IV in Barnes et al. (1968), or the new classes of Data Base computers in Hsiao (1980) and Data Flow computers in Davis and Drongowski (1980).

2. In normal practice, the informal design phase involves the making of a systematic series of decisions leading to the conceptual design of each component and the interrelationships among components. The decision process includes the use of empirical data, the selection of appropriate architectural styles, the adaptation or instantiation of known schemata for particular components, and the direct adoption of known architectural principles. Thus the lesser but more usual form of creativity involves innovation in adaptation or adoption of concepts from an established corpus of knowledge to a relatively new design situation.

Example 14.1

The design of the control unit for the Motorola MC68000 reported in Stritter and Tredennick (1978) does not involve any new architectural principle, but rather the application of a previously developed concept—a two-level control store—as a means to cope with the problem of designing a single-chip 16-bit processor.

To resume the QM-C example, in designing its storage organization it was decided to implement a stack in the QM-C's main memory to hold local variables and other dynamic objects. As discussed in Chapter 13, one of the most important aspects of language-directed architectural design is the implementation of efficient procedure call and return mechanisms. Since the early days of Algol 60 (Randell and Russell [1964]) the procedural or control stack has emerged as a most elegant schema for storage management during

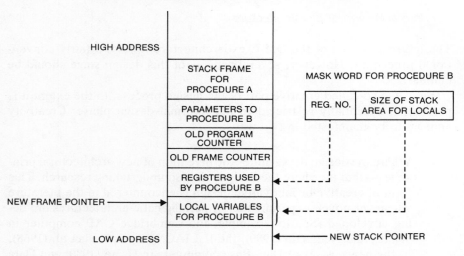

Figure 14.5 State of the QM-C stack after a procedure call.

entry to and exit from procedures and blocks. The design of the stack frame, then, represents a direct instantiation of this schema, in that the general schematic notion of the stack frame was adopted and suitably modified to fit the context of the QM-C.

The QM-C stack will also be the subject of later discussions of the formal design phase, so it is appropriate to provide a full description of its informal design; see Olafsson (1981, pp. 43–44).

The stack is used to hold local variables, parameters for procedures, and the contents of registers when context switching occurs. The calling procedure begins the call by pushing arguments to the called procedures onto the stack. The **call** instruction is then executed. After calculating the effective address of the called procedure, its *mask word* is read. This is the first word of the called procedure and contains the lowest register number used by the called procedure along with the size of the stack area needed for local variables. On the basis of this information, registers are saved and storage for automatics (local variables) is allocated on the stack. Finally, control is transferred to the first instruction of the called procedure. Figure 14.5 shows the state of the top of the stack just after a procedure A has called a procedure B with two parameters.

In returning from the procedure essentially the reverse operations take place; the return instruction has as its operands the register number and the size of the argument list. The registers are popped from the stack, including the old frame pointer. The last thing done before control is returned to the calling procedure is the deallocation of the argument area from the stack.

As one more example of the modus operandi of informal design, consider the method of selecting the order code. The following passages are taken from Olafsson (1981, pp. 53–54).

> The method of selecting operators [operations] is derived directly from Hoevel's work on Directly Executable Languages (DEL's) [Hoevel 1978, 1974].
>
> Assume that we have a set $F = \{f(1), f(2), \ldots, f(n)\}$ of potential operators. The task is to decide which of the operators f(i) should be candidates for direct implementation as nanocoded instructions $x(i)$ and which should be implemented entirely using other operators $\{f(j), j \neq i\}$ applying the idea of *macro-expansion* X(i) of f(i). We wish to determine a subset G of F that in some sense minimizes a significant percentage of the code. We chose to select members of G on the basis of our study on the C code base. The task now becomes simply to select the most common operators in C in order of their frequency until either we run out of nanostore or have exhausted the set F. In the light of the limited opcode space it is clear that we will run out of opcode space long before F is exhausted. The remaining operators in F must then be implemented with macro-expansion using the already selected operators.

Thus, we see that the order code design was influenced on the one hand by the use-statistics for C (cf. Chapter 13) and on the other by Hoevel's work on DELs. Another direct factor contributing to this design is the concept of *reduced instruction sets* proposed recently by Patterson and his coworkers in Patterson and Ditzel (1980) and Patterson and Sequin (1981)—the basic notion being that where resources for implementation of instruction sets are scarce (as in VLSI single-chip computers), they may best be used by reducing the complexity and size of the order code directly supported by the hardware and firmware (see also Section 12.2). Thus, in the case of the QM-C, Olafsson (1981, p. 54) explains:

> Rather than reserving an opcode for the direct implementation of [higher level] statements like *for* loops, it was decided that macro-expansion would take care of their implementation and use the opcode thus freed to provide size or speed (or both) optimizations to the more frequent (simple) instructions. (For example, adding a constant to a register is performed at least 6 times more often than *for* loops; thus using the freed opcode to provide a one word optimization of this instruction would result in more savings than an instruction supporting the execution of *for* loops.)
>
> One of the effects of this method of selecting operators is that the resulting instruction set becomes simpler and closer to . . . a *Reduced Instruction Set* (RIS) . . . . Another argument for selecting a RIS for the QM-C is the pipelined nature of the [QM-1] instruction

execution mechanism; the time taken to execute n one-word instructions (in a straight line segment) can be essentially the same as executing one n-word instruction. Thus, if the macro-expansion of one n-word instruction is n one-word instructions nothing is gained but one precious opcode is "wasted."

It is not necessary to go any further into the details of the QM-C exoarchitecture. The main features are summarized in Section 14.3.2 below, while for a thorough discussion the reader is referred to Olafsson (1981). The important points to note are the extensive applications of empirical data in the decision making process, adoption of established principles, adaptation and instantiation of schemata, and invention of new architectural principles as main components of the informal design stage. We also saw that the partitioning of the QM-C exoarchitecture design problem was along natural lines set by the view of the machine as seen by a potential user.

14.3.2 Informal Design of the QM-C Endoarchitecture

Recall from an earlier discussion in Section 14.3 that the overall design style selected was the so-called outside-in approach. That is, the ordering of the design decisions was intended to proceed from the exoarchitectural level through endoarchitecture to the implementation in nanocode (Fig. 14.2). Thus far I have only discussed the general nature of the QM-C exoarchitecture as it emerged through informal design.

The architect has a few choices at this point. For instance he or she may complete the formal design of exoarchitecture and then proceed to the endoarchitectural level. Alternatively, the architect may consider the informal design of the endoarchitecture first, before proceeding with the formal design of both exoarchitecture and endoarchitecture.

We have no evidence as yet of which of these, or any other alternative, is the better approach. However, given our emerging understanding of the design process, one may surmise that each approach is likely to be more appropriate in specific circumstances. The former, for example, would be the appropriate choice when there is a clear separation of exoarchitectural and endoarchitectural designs, as is the case when an essentially implementation-independent exoarchitecture is being designed. See for example discussions of the SWARD exoarchitecture in Myers (1982) or of the Concurrent Pascal machine in Brinch Hansen (1977).

However, in the case of the QM-C, where the concern is the design of an integrated architecture that is intimately related to a particular implementation style, it seems more useful and productive to develop the informal design of both architectures before proceeding to the formal design phase.

In designing the QM-C, much of the endoarchitecture already exists in the form of the QM-1 nanolevel architecture. The informal design of the endoarchitecture then essentially involves two tasks: the mapping of the QM-C exoarchitecture onto the QM-1 nanoarchitecture—that is, deciding how to represent exoarchitectural features on the QM-1—and formulating the QM-C instruction interpretation mechanism in terms of the available emulation capabilities of the QM-1.

As we have already seen, much of the first task was done during the informal design of the exoarchitecture, since the latter was influenced by the QM-1. Further, certain gross aspects of the instruction interpretation mechanism were developed hand-in-hand with the exoarchitectural design, since this also determined some key decisions concerning the exoarchitecture. In particular, the general structure of the instruction decoding mechanism was determined during the informal stage of exoarchitecture design (Fig. 14.3; see also the remarks on choice of implementation style in Section 14.3.1 above).

To complete this discussion it would be useful at this stage to summarize the salient features of the QM-C architecture as it appears at the end of the informal design phase. These may be listed as follows:

1. An 18-bit word main store mapped onto the QM-1 control store.
2. Three distinct classes of data objects: global variables that reside in QM-C main store; local variables that are allocated storage in the stack frame for each procedure invocation; and registers. The QM-C register file consists of eight registers used to hold variables, four for expression evaluation and other temporary storage, and five dedicated to special purpose functions (e.g., the program counter and the stack pointer). All QM-C registers are 18 bits wide and are mapped onto the QM-1 local store.
3. A control (procedure) stack, resident in QM-C main memory and used for dynamic storage allocation during procedure invocation.
4. Three different addressing modes: (a) "register + offset" for global, local and parameter variables; (b) "register" for register variables; and (c) "immediate" for constant operands.
5. Data types supported by the QM-C include the "primitive" C data types integer, character, long (integer), and short (integer)—all represented by 18-bit words—and the composite data types array, pointer, string, and structure.
6. Six basic formats for two-operand instructions that use the three addressing modes cited in (4) above:
    a. CM: constant-to- (local or global) memory variable.
    b. CR: constant-to-register.
    c. RR: register-to-register.
    d. RM: register-to-memory.
    e. MR: memory-to-register.
    f. MM: memory-to-memory.
7. An order code (instruction set), determined primarily from the empirical study of C programs (see Chapter 13 and the comments on map-

ping the outer environment in Section 14.3.1 above), which may be partitioned into Flynn's (1974) three classes—functional (F-type), procedural (P-type), and move (M-type). The F-type class includes such operations as increment/decrement, add/subtract, and/or, left and right arithmetic shift, and multiplication/division. The P-type instructions include branches and the procedure call and return operations. The two main subclasses of the M-type set are MOV and PUSH—the former responsible for data movement involving the register file and main memory and the latter placing parameters on the stack before procedure calls. All opcodes are seven bits wide.

8. Finally, an instruction execution cycle of the following nature:
    a. Instruction fetch: control store is read as 18-bit words and the beginning and end of instructions located.
    b. Instruction decode: operands are located in the instruction, a nanostore address generated, and control transformed to the nanoroutine interpreting the instruction.
    c. Instruction execution: effective addresses of operands are calculated and the instruction executed.

As this summary suggests, most of the detailed endoarchitecture (which, of course, reflects the exoarchitecture) actually emerged during the formal design phase. It is to this we now turn our attention.

14.3.3 Formal Design

Formal design, as we recall, has several purposes: to enhance the reader's understanding of the architecture; to raise the designer's confidence as to its correctness; to provide a basis for rigorous validation, evaluation, and testing of the architecture before its implementation; and finally, to document a precise specification of the architecture.

The heart of this design phase is a description in some formal language. In the case of the QM-C this was done by using S*A. The resulting architectural description was then refined and transformed into a lower-level emulation program specified in S*(QM-1) (see Chapter 11). The emulator is intended to run at the nanoarchitectural level of the QM-1, with the object program residing in control store (Fig. 14.6).

Clearly, given the stated objectives of formal design, the decisions to be made and the factors that must be entertained during this design phase may differ considerably—even radically—from those attending the informal design. The entire formal design process must be informed by the desire for clarity, simplicity, and ease of understanding, whereas the primary concerns during informal design were functionality, economy, and implementability.

This, in fact, raises a serious problem: in the model of the design process proposed here, I have indicated a clear separation of the informal and formal aspects of design. It is thus quite possible that the informal design—

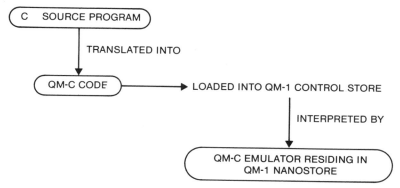

Figure 14.6 Emulation scheme in QM-C.

optimized for functionality, economy, and implementability—may not be the most appropriate from the viewpoint of formal design. For instance it may be difficult to formally describe the architecture, and still more difficult to verify design correctness. The objectives of informal and formal design may turn out to be mutually contradictory.

There are no easy solutions for this dilemma. Two possibilities suggest themselves, however. First, just as in writing the specifications of a software system before its construction, as little commitment as possible is made to implementation details (Parnas [1972]), so in designing architectures the focus, during the informal design phase, should be on policy decisions, leaving it to the formal design stage to devise the mechanisms that realize these policies. Second, design is inevitably an iterative process. Once the informal/formal design sequence has been attempted or completed, the cycle can be repeated with suitable feedback of the understanding and insight gained in the preceding iteration.

In formally describing the QM-C in S*A, all the information accumulated in the informal design is embedded in a procedural specification of the architecture. That is, the QM-C architecture is viewed as a system of interacting processes, and the structure and behavior of this system define the architecture.

Space does not permit me to provide even a reasonably complete S*A description of the QM-C. Nor would it be particularly useful to do so. Thus I shall restrict the formal description to a few small fragments from the complete S*A specification.

Example 14.2

The overall composition of the QM-C system is shown in Fig. 14.7. The system is partitioned into four simpler systems named INSTRUCTION CYCLE, MEMORY, INTERRUPT, and NANO_ARCHITECTURE. IN-

```
sys QM_C;

    sys INSTRUCTION_CYCLE;

        mech INSTRUCTION_FETCH;   ... endmech;
        mech INSTRUCTION_DECODE;  ... endmech;

        sys INSTRUCTION_EXEC;

            mech FETCH_OPRND;        ... endmech;
            mech EXECUTE;            ... endmech;
            mech DEPOSIT_OPRND;      ... endmech;

        endsys;

    endsys;

    sys MEMORY;

        mech MAIN_MEM;  ... endmech;
        mech AUX_MEM;   ... endmech;

    endsys;

    sys INTERRUPT;

        mech INT_HANDLERS;   ... endmech;
        mech STATE_CHANGE;   ... endmech;

    endsys;

    sys NANO_ARCHITECTURE;

        mech EXEC_PRIM;  ... endmech;
        mech IO_PRIM;    ... endmech;
        mech MEM_PRIM;   ... endmech;

    endsys;

endsys;
```

Figure 14.7 Overall composition of the QM-C system (copyright © 1982, IEEE; reprinted with permission).

STRUCTION_CYCLE consists of two mechanisms, INSTRUCTION FETCH and INSTRUCTION_DECODE, and a subsystem named INSTRUCTION_EXEC. The remaining principal subsystems are also composed of sets of mechanisms.

Example 14.3

A portion of the INSTRUCTION_CYCLE system is shown in Fig. 14.8. It contains a set of data type and data object declarations relevant to this and subsequent examples. The INSTRUCTION_FETCH mechanism contains a single public procedure that activates the main memory mechanism to prefetch an instruction, while the INSTRUCTION_DECODE mechanism decodes the previously fetched instruction.

DESIGN OF THE QM-C ARCHITECTURE

```
sys INSTRUCTION_CYCLE;

    type ls_register = seq[17..0]bit;
    type f_register  = seq[5..0]bit;
    type bus         = seq[17..0]bit;
    .
    glovar main_store_output : bus;
    glovar local_store : array[0..31]of ls_register
                         with fmod, fcod, faod, feod, fair, fail, gspec
                       : tuple
                             general_purpose  : array[0..23]of ls_register
                             index            : array[0..3]of ls_register
                                                with fmpc
                             general_purpose2 : array[0..2]of ls_register
                             inst_reg         : ls_register
                         endtup;
    glovar f_store     : array[0..31]of f_register
                       : tuple
                             fmix, fmod, fcia, fail, fcid, fair : f_register
                             fcod, faod, fsid, fsod, feid, feod : f_register
                             feia, feoa, fact, fusr, fmpc, fidx : f_register
                             fist, fiph,                        : f_register
                             backup               : array[0..11]of f_register
                         endtup;
    .
    const n_pc  : dec(6) 25;
    const n_sp  : dec(6) 28;
    const n_fp  : dec(6) 24;
    const n_eb  : dec(6) 26;
    const n_eb1 : dec(6) 02;

    mech INSTRUCTION_FETCH;
    /*
     * Context of Activation : at end of each instruction execution,
     * and interrupt handlers.
     *
     * Decoding of a previously fetched instruction (say i) is initiated
     * in parallel with the fetching of the new instruction (i+1).
     *
     */
      proc NEXT;

          act MAIN_MEM.READ_P1   act INSTRUCTION_DECODE.SELECT_NEXT;

      endproc;
    endmech;
    . . . .
    . . . .
endsys;
```

Figure 14.8 A partial description of the QM-C instruction cycle (copyright © 1982, IEEE; reprinted with permission).

In Chapter 1 I suggested that a machine's exoarchitecture is basically an abstraction of its endoarchitecture: certain aspects of the latter are deliberately suppressed since they play no part in the user's understanding of the computer. Thus, the QM-C exoarchitecture would be formally represented by the following S*A system:

sys QM-C; /* exoarchitecture */
        sys INSTRUCTION_EXEC;
            mech FETCH_OPRND; . . . . . . **endmech**;

```
        mech EXECUTE; . . . . . . . . . . endmech;
        mech DEPOSIT_OPRND; . . . . endmech;
     endsys;

        sys MEMORY;
           mech MAIN_MEM; . . . . endmech;
           mech AUX_MEM; . . . . . endmech;
        endsys;

        sys INTERRUPT; . . . . endsys
     endsys
```

In essence, the instruction fetch and decode mechanisms, the NANO_ARCHITECTURE system, and possibly a part of the INTERRUPT system are omitted in the exoarchitectural specification.

As a more detailed example of formal design, we consider here the description of the QM-C CALL instruction introduced earlier in Section 14.3.1. Recall that the purpose of this instruction is to save the contents of

```
syn main_output = control_store_output;          /* Main memory output bus */
syn reg : array[0..31]of ls_register             /* QM-C register file    */
             with dest, source, addr, alt_source,
                  inst_reg.ab.a, inst_reg.ab.b
           : tuple
               dum   : array[0..11]of ls_register
               temp  : array[0..3]of ls_register  /* Temporaries    */
               var   : array[0..7]of ls_register  /* Variable reg.  */
               index : array[0..3]of ls_register
                       with mm_index_select
                     : tuple
                         fp : ls_register         /* Frame pointer     */
                         pc : ls_register         /* Program counter   */
                         eb : ls_register         /* External base     */
                         ax : ls_register         /* Aux mem index     */
                       endtup
               sp    : ls_register                /* Stack pointer     */
               scr1  : ls_register                /* Nano scratch reg 1 */
               scr2  : ls_register                /* Nano scratch reg 2 */
               inst_reg : tuple
                            opcode : seq[6..0]bit
                            ab     : tuple
                                       a : seq[4..0]bit
                                       b : seq[5..0]bit
                                     endtup
                          endtup
                        : tuple                   /* 6 // 18 interface */
                            c : f_register
                            a : f_register
                            b : f_register
                          endtup
             endtup = local_store;
syn mm_addr_select  : seq[5..0]bit   /* Selects main memory address source */
                    = f_store.fcia[5..0];
syn mm_index_select : seq[1..0]bit   /* Selects addr+offset source         */
                    = f_store.fmpc[1..0];
syn mm_data_select  : seq[5..0]bit   /* Selects mm data source             */
                    = f_store.fcid[5..0];
```

Figure 14.9 Synonym declarations (copyright © 1982, IEEE; reprinted with permission).

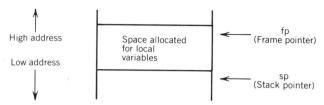

Figure 14.10 State of the QM-C stack before a procedure call (copyright © 1982, IEEE; reprinted with permission).

the QM-C registers and allocate space on a stack before transferring control to the called procedure. With reference to the overall description of the QM-C, CALL is defined as one of the global procedures inside the EXECUTE mechanism, which in turn is part of the INSTRUCTION_EXEC system (Fig. 14.7). The data objects used by CALL are defined as synonyms of previously defined global variables (Fig. 14.8) and are shown in Fig. 14.9.

Just before a CALL instruction is executed, the stack is as shown in Fig. 14.10; the first word of the procedure being called is a mask whose format is shown in Fig. 14.11 (see the discussion of informal design of exoarchitecture in Section 14.3.1 for an explanation for this mask). In addition the following preconditions are assumed to hold: (1) fp points to the start of the activation record for the calling procedure; (2) pc points to the CALL instruction in memory; (3) eb denotes the base address for the entire C object program; (4) inst_reg. contains the operand of the CALL instruction (i.e., it is an 11-bit constant that points to a word in memory relative to the base address specified in eb that holds the aforementioned mask).

Given these conditions, the sequence of actions resulting from the CALL instruction is given by the S*A procedure in Fig. 14.12. This procedure includes calls to three (of eleven) public procedures within the mechanism MAIN_MEM, and to the INSTRUCTION_FETCH mechanism (see Fig. 14.8). The relevant portion of MAIN_MEM is shown in Fig. 14.13.

The next stage is to transform or translate this architectural description into a microprogram. Because of the kinship between S*A and S* (and therefore between S*A and S*[QM-1]), the microprogram will be very similar to the architectural description, with the following important differences:

1. The constructs **mech** and **sys,** which are used in S*A for global structuring of architectural descriptions, are too high-level for specification of microprograms. Further, in S*(QM-1), procedures are func-

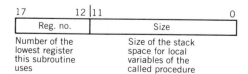

Figure 14.11 Mask format (copyright © 1982, IEEE; reprinted with permission).

```
proc CALL;
    do reg.index.pc := reg.index.pc+1        /* Point pc to next instruction */
       reg.scrl := reg.inst reg              /* Save offset from inst.       */
       mm_index_select := n_eb
    od;

    call MAIN_MEM.READ_I_L;                  /* read mask work onto bus      */
    reg.inst_reg := main_output;             /* Prepare to decode            */

    do mm_addr_select := n_sp                /* Prepare to save reg's        */
       mm_data_select := n_pc+1
    od;

    repeat                                   /* Save registers on stack,     */
        mm_data_select := mm_data_select - 1 /* iteratively until the register*/
        call MAIN_MEM.PUSH.G                 /* specified in inst_reg.c has  */
    until mm_data_select = reg.inst_reg.c;   /* been saved                   */

    reg.index.fp := reg.sp;                      /* Set new frame pointer    */
    reg.sp := reg.sp - reg.inst_reg.ab;          /* Allocate for automatics  */
    reg.index.pc := reg.index.eb + reg.scrl+1;   /* First instruction of proc*/

    mm_index_select := n_pc;
    call MAIN_MEM.READ_I;                    /* Read onto bus for decode     */

    act INSTRUCTION_FETCH.NEXT;              /* Activate instr. pipeline     */
endproc;
```

Figure 14.12 Description of the CALL instruction in S*A (copyright © 1982, IEEE; reprinted with permission).

tionally distinguished as implementing subroutines, interrupts, and instructions. Thus, the transformation from S*A to S*(QM-1) requires reorganizing the original description from the **sys/mech/proc** hierarchy to a set of **proc** and **macro** statements in S*(QM-1).

2. Generally speaking, the data objects specified in an S*A description are precisely those that contribute to the particular exoarchitecture or endoarchitecture being designed. These data objects may be global or local to particular mechanisms. In S* and its instantiations, data objects are global to all procedures, and represent actual memory resources in the host machine. Thus in transforming the description from S*A into S*(QM-1), appropriate mappings of data objects must be performed. This task is greatly facilitated by the fact that the same data types exist in both languages.

   In designing and specifying the QM-C data objects (Fig. 14.8), advantage was taken of the fact that this architecture would be emulated on the QM-1. Thus, the data object declarations in S*A are almost identical to the (predefined) declarations of variables and constants in S*(QM-1). The mapping, in this case, was trivial.

3. The body of an individual procedure in an S*A description is a specification of the control of information flow and its transformation in the proposed machine. The procedure in the corresponding S*(QM-1) description is a specification of a program, that is, a symbolic representation of the contents of nanostore whose systematic execution will cause the appropriate gates to be controlled in the QM-1. Thus the major difference between the S*A description and the

```
mech MAIN_MEM;
  /* Controls access to main memory. Activated by memory accessing instructions
   * and inst. fetch mechanism.
   * Public procedures :
   * . . . . . .
   * READ_I_L - reads MM at address in index reg. + 'ab' field of inst. reg.
   * READ_I  - reads MM of address in index reg.
   * PUSH_G  - writes MM at address in gen.reg; reg. decrmtd by 1
   * after write */

  privar main_mem : array[0..40959]of seq[17..0]bit with
                                                    mm_addr_so;

  proc READ_I;
    /* read main mem; address from fp, eb or pc; mm_index_select
     * must be set by caller to select any of these registers */
    main_output :=
        main_mem[mm_addr_so.index[mm_index_select]]
  endproc;

  proc READ_I_L;
    /* read main mem; address formed by adding the ab field of
     * inst_reg to the index reg. (fp, eb, or pc) selected by
     * mm_index_select */
    main_output :=
        main_mem[mm_addr_so.index[mm_index_select]+reg.inst_reg.ab]
  endproc

  proc PUSH_G;
    /* write main_mem; address in gen. reg; mm_addr_select must
     * be set by caller to select register; mm_data_select selects
     * data source. Reg. decremented by 1 after write. Note that
     * data written also appears on the main mem output bus */
    do
      main_mem[mm_addr_so.reg[mm_addr_select]] :=
            mm_data_so[mm_data_select]
      main_output := mm_data_so[mm_data_select]
    od;
    mm_addr_so.reg[mm_addr_select] :=
        mm_addr_so.reg[mm_addr_select] - 1
  endproc;
  . . . . . . . . .
endmech
```

Figure 14.13 MAIN_MEM mechanism.

S*(QM-1) implementation lies in the operations defined on the data objects, as opposed to the data objects themselves. For instance, an individual assignment statement in S*A may have to be mapped into one or more machine-specific statements in S*(QM-1).

For example, the statement

$$\text{reg.index.pc} := \text{reg.index.eb} + \text{reg.scr1} + 1$$

in Fig. 14.12 is represented in S*(QM-1) by the sequence of statements:

$$\text{alu\_carry\_in} := 1;$$
$$\text{pc} := \text{eb} + \text{scr1};$$

(In S*(QM-1), a field within a tuple variable may be referenced without prefixing the field name with the variable identifier as long as the field name is unique.)

```
proc p_call (instruction, mnemonic=call)
    /*
     * Macros : INC_PROC_COUNT        - Add one to program counter
     *          MAIN_MEM_READ_I_L     - Read memory (base plus long offset)
     *          MAIN_MEM_READ_I       - Read memory address in index register
     *          MAIN_MEM_PUSH_G       - Push onto stack
     *          INSTRUCTION_FETCH_NEXT- Prefetch (program counter not updated)
     *
     */

    INC_PROG_COUNT;                          /* Point pc to next instruction */
    scr1 := passl inst_reg;                  /* Save on stack                */

    cocycle
        mm_index_select := c_eb              /* Read from external base      */
        mm_addr_select  := c_sp              /* Save on stack                */
        mm_data_select  := C-pc_pl           /* PC first saved (c_pc+1)      */
    coend;

    MAIN_MEM_READ_I_L;                       /* Read mask word               */
    inst_reg := main_output;                 /* Prepare to decode            */

    repeat                                   /* Save registers               */
      mm_data_select := mm_data_select - 1;
      MAIN_MEM_PUSH_G
    until (mm_data_select = inst_reg.c);

    inst_reg.c := 0;                         /* Prep. to allocate automatics */
    fp := passl sp;                          /* Set new frame pointer        */
    sp := sp - inst_reg;                     /* Allocate for automatics      */

    alu_carry_in := 1;                       /* Add one to following expr.   */
    pc := eb + scr1;                         /* First instruction of proc    */

    mm_index_select := c_pc;
    MAIN_MEM_READ_I;                         /* Read onto bus for decode     */

    INSTRUCTION_FETCH_NEXT;                  /* Activate instr. pipeline     */

endproc
```

Figure 14.14 Description of the CALL instruction in S*(QM-1) (copyright © 1982, IEEE; reprinted with permission).

The microprocedure (of the class *instruction*) for the QM-C CALL instruction is defined in S*(QM-1) as shown in Fig. 14.14. The declaratives are almost identical to those defined earlier and have therefore been omitted.

The principal conceptual differences between Figs. 14.12 and 14.14 are that (1) the QM-C's main memory is represented by the QM-1's control store and (2) a **cocycle** . . . **coend** statement appears in the microprogram for the purpose of specifying the execution of a machine-specific parallel statement. This particular **cocycle** statement indicates that its three constituent assignment statements will begin and complete execution in parallel in the same microcycle (which, in the QM-1, corresponds to a single T-step; see Nanodata [1979]). And finally, (3) the mechanism calls in the S*A description are replaced by macro calls in the S*(QM-1) version.

14.3.4 Verification of Architecture

One of the main aims of formal design is to describe the design so that it may, at least in principle, be formally verified or tested (using simulation) before implementation. Let us consider this issue in the context of the present example.

Suppose that we have complete formal descriptions of the QM-C in both S*A and S*(QM-1): the former describes the architecture of the QM-C machine, the latter, the emulator intended to implement the architecture on a given host machine, the QM-1. Thus the problem of correctness arises in both contexts.

Consider the CALL instruction. At first sight, on comparing Fig. 14.12 and 14.14 it would appear that the latter is not much more complex than the former, so that the difficulty of verification would be roughly of the same order in both cases.

In fact, this is not the case. Verification of the S*(QM-1) microprogram will be an order of magnitude more difficult because many of the proof rules for S*(QM-1) are considerably more complex than those for S*A. That is, the semantics of an S*(QM-1) construct is more complex than the semantics of a similar construct in S*A.

In this chapter, we will not be considering the proof of correctness of the S*(QM-1) program. The interested reader may refer to Wagner (1983) and Wagner and Dasgupta (1983) for detailed discussions of this topic.

Remarks on Notation

1. In Fig. 14.5, consider the stack pointer; this is represented in the QM-C by "reg.sp," which is a data object of type "ls_register." In developing the assertions we shall find it convenient (and more understandable) to tag a data object identifier with its type. For example, the stack pointer will be designated within assertions as "reg.sp: ls_register."

2. We have also seen that through the use of synonyms a data object of a given type with a given identifier may be alternatively viewed as a data object with a different name of some other type. For instance, in the QM-C description there exists a synonym declaration (not shown in Fig. 14.9):

    syn mm_addr_so: /* main mem address source */
      tuple
        reg : **array** [0..31] **of** ls_register
                    with mm_addr_select;
        xx : bus;
        index : **array** [0..3] **of** ls_register
                    with mm_index_select;
        yy : bus
      endtup = control_store_address_source

    where "control_store_address_source" is a variable declared as:

glovar control_store_address_source:
  tuple
    local_store : **array** [0..31]
              of ls_register **with** fcia
    control_store_output : bus
    index : **array** [0..3] **of** ls_register **with** fmpc
    index_alu_output : bus
  endtup

Now, since "n_sp" is a name for the constant "28" (see Fig. 14.8) and local_store [28] in the QM-1 is used as the stack pointer, the latter is not only denoted by the symbol "reg.sp : ls_register," but also by "mm_addr_so.reg [mm_addr_select] : ls_register" and by "mm_addr_so.reg [n_sp] : ls_register." In specifying assertions involving a data object with alternative identifiers X, Y, . . . , Z, I shall also, where necessary, use the notation "X | Y | . . . | Z" meaning "X or Y or . . . or Z."

3. Finally, in order to simplify the assertions and enhance their readability, auxiliary variables will be introduced wherever necessary. Auxiliary variables appear in the assertions but not in the text. These will be denoted by subscripted identifiers, such as x_0, y_1, and so on.

The key to verification is, as we have seen in earlier chapters, the construction of appropriate assertions that capture precisely the intended state of the machine at specific points. We shall accordingly construct these assertions for the CALL description, insert them into the S*A text, and obtain a proof outline.

The CALL procedure invokes three MAIN_MEM procedures. Proof outlines for these are first given below:

1. Let PRE_READ_I_L denote the assertion:

    mm_index_select : **seq**[1..0]**bit** = x_0 &
    reg.inst_reg.ab : **seq**[10..0]**bit** = i_0 &
    mm_addr_so.index[x_0] : ls_register = b_0

Then the proof outline for the READ_I_L procedure is simply:

    {PRE_READ_I_L}

main_output := main_mem[mm_addr_so.index[mm_index_select]
              + reg.inst_reg.ab]

    {main_output : bus = main_mem[b_0 + i_0] :
        seq[17..0]**bit** & PRE_READ_I_L}

DESIGN OF THE QM-C ARCHITECTURE

2. Let PRE_PUSH_G denote the assertion:

   mm_addr_select : **seq[5..0]bit** = r_0 &
   mm_data_select : **seq[5..0]bit** = t_0 &
   mm_addr_so.reg[r_0] : ls_register = a_0 &
   mm_data_so[t_0] : ls_register | reg[t_0] :
          ls_register = d_0

Then, for PUSH_G we have:

 {PRE_PUSH_G}

do
 main_mem[mm_addr_so.reg[mm_addr_select]]
   := mm_data_so[mm_data_select]
☐ main_output := mm_data_so[mm_data_select]
od

{INT_PUSH_G ≡
 main_mem[a_0] : **seq[17..0] bit** = d_0 &
 main_output : bus = d_0 & PRE_PUSH_G}

mm_addr_so.reg[mm_addr_select]
  := mm_addr_so.reg[mm_addr_select] − 1

{POST_PUSH_G ≡
 main_mem[a_0] : **seq[17..0] bit** =
   d_0 & main_output : bus = d_0
& mm_address_select : **seq[5..0] bit** = r_0
& mm_data_select : **seq[5..0] bit** = t_0
& mm_data_so[t_0] : ls_register = d_0
& mm_addr_so.reg[r_0] : ls_register = a_0 − 1}

3. Finally, let PRE_READ_I denote the assertion:

   mm_index_select : **seq[1..0] bit** = x_0 &
   mm_addr_so.index [x_0] : ls_register = b_0

Then, for READ_I we have the proof outline:

 {PRE_READ_I}

main_output := main_mem[mm_addr_so.index[mm_index_select]]

 {main_output : bus = main_mem[b_0] : **seq[17..0] bit** &
   PRE_READ_I}

Consider now the procedure for the CALL instruction. At the time it begins execution (see discussion in Section 14.3.3) the following assertion holds:

PRE_CALL:
 reg.index.fp : ls_register = fp_0
& (reg.sp : ls_register|mm_addr_so[n_sp] :
     ls_register = sp_0)
& (reg.index.pc : ls_register|mm_data_source[n_pc] :
     ls_register = pc_0)
& (reg.index.eb : ls_register|mm_addr_so.index[n_eb] :
     ls_register = b_0)
& (reg.inst_reg : ls_register = opd_addr_0
& main_mem[b_0 + opd_addr_0] [17..11] = r_lowest_0 :
     16 ≤ integer ≤ 23
& main_mem[b_0 + opd_addr_0] [10..0] = $local\_space_0$: integer
& ($\forall j$: 23, 22, . . . , r_lowest_0) (reg.var[j] = $reg.var_0[j]$)

The intended postcondition of the CALL instruction before activating INSTRUCTION_FETCH.NEXT will be such that:

1. The state of the stack and associated pointers should be as shown in Fig. 14.5. Note that the saved value of the program counter pc (which was previously pointing to the CALL instruction itself) must be such that it points to the instruction following CALL (in the calling procedure).
2. The program counter is pointing to the first instruction of the called procedure.
3. The first instruction of the called procedure is on the main memory output bus.

We may formally describe this postcondition by means of the assertion:

POST_CALL:
 main_mem[sp_0] = pc_0 + 1
& main_mem[sp_0 − 1] = fp_0
& ($\forall j$: 23, . . . , r_lowest_0) (main_mem[sp_0 − (25 − j)] =
     $reg.var_0[j]$)
& reg.index.fp : ls_register = sp_0 − (25 − r_lowest_0) − 1
& reg.sp : ls_register|mm_addr_so[n_sp] : ls_register =
  sp_0 − (25 − r_lowest_0) − 1 − $local\_space_0$
& reg.index.pc : ls_register|mm_data_source[n_pc] :
     ls_register = b_0 + opd_addr_0 + 1
& main_output : bus = main_mem[reg.index.pc]

The procedure CALL achieves the transformation from PRE_CALL to POST_CALL in the following stages:

1. Starting with PRE_CALL it first effects a state change such that the following intermediate assertion holds:

    INT_CALL1:
    reg.index.eb : ls_register | mm_addr_so.index[n_eb] :
    $\qquad\qquad\qquad$ ls_register = b_0
    & reg.scr1 : ls_register = opd_addr$_0$
    & reg.inst_reg.c = r_lowest$_0$
    & reg.inst_reg.ab = local_space$_0$
    & main_mem[sp$_0$] = pc$_0$ + 1
    & main_mem[sp$_0$ - 1] = fp$_0$
    & $\forall j$: 23, ..., r_lowest$_0$) (main_mem[sp$_0$ - (25 - j)] =
    $\qquad\qquad\qquad$ reg.var$_0$[j])
    & reg.sp : ls_register | mm_addr_so[n_sp] :
    $\qquad\qquad$ ls_register = sp$_0$ - (25 - r_lowest$_0$) - 1
    & reg.index.fp : ls_register = fp$_0$

    Informally, INT_CALL1 reflects the state at which the values of the (incremented) program counter, the frame pointer, and the QM-C general registers have been pushed onto the stack, the previous contents of the instruction register saved in a scratch register, and the mask made available in the instruction register.

2. Given that INT_CALL1 holds, the procedure next causes a state change such that the following assertion holds:

    INT_CALL2:
    $\qquad$reg.index.eb : ls_register | mm_addr_so.index[n_eb]
    $\qquad$: ls_register = b_0
    & reg.inst_reg : ls_register = opd_addr$_0$
    & main_mem[sp$_0$] = pc$_0$ + 1
    & main_mem[sp$_0$ - 1] = fp$_0$
    & ($\forall j$: 23, ..., r_lowest$_0$) (main_mem[sp$_0$ - (25 - j)] =
    $\qquad\qquad\qquad$ reg_var$_0$[j])
    & reg.sp : ls_register | mm_addr_so[n_sp] : ls_register =
    $\qquad\qquad$ sp$_0$ - (25 - r_lowest$_0$) - local_space$_0$
    & reg.index.fp : ls_register = sp$_0$ -
    $\qquad\qquad\qquad$ (25 - r_lowest$_0$) - 1

    That is, space for the local variables in the called procedure has been allocated on the stack and the frame pointer is pointing to the starting position of this space.

```
proc CALL;
  {PRE_CALL}
  do reg.index.pc := reg.index.pc + 1
     reg.scr1 := reg.inst_reg
     mm_index_select := n_ebl
  od;
  call MAIN_MEM.READ.I_L;
  reg.inst_reg := main_output;
  do
     mm_addr_select := n_sp
     mm_data_select := n_pc + 1
  od;
  repeat
     mm_data_select := mm_data_select - 1;
     call MAIN_MEM.PUSH_G
  until mm_data_select = reg.inst_reg.c
  {INT_CALL1}
  reg.index.fp := reg.sp;
  reg.sp := reg.sp - reg.inst_reg.ab;
  {INT_CALL2}
  reg.index.pc := reg.index.eb + reg.scr1 + 1;
  mm_index_select := n_pc;
  call MAIN_MEM.READ_I;
  {POST_CALL}
  act INSTRUCTION_FETCH.NEXT
endproc
```

Figure 14.15 Proof outline for the CALL instruction.

3. Finally, the rest of the procedure is responsible for transforming the machine to a state such that the assertion POST_CALL is true.

The complete proof outline is shown in Fig. 14.15; most of this proof is quite trivial and I shall therefore only give details of the correctness of the iterative statement.

It can be easily verified that before the start of the loop, the following assertion holds:

PRE_LOOP:
$\quad$ (reg.index.eb : ls_register | mm_addr_so.index[n_eb]:
$\qquad\qquad\qquad$ ls_register = b_0)
& reg.scr1 : ls_register = opd_addr_0
& reg.inst_reg.c = r_lowest_0
& reg.inst_reg.ab = $local\_space_0$
& reg.index.fp : ls_register = fp_0
& reg.sp : ls_register | mm_addr_so[n_sp] : ls_register = sp_0
& mm_addr_select : **seq**[5..0]**bit** = 28
& mm_data_select : **seq**[5..0]**bit** = 26

When the **repeat** statement terminates, INT_CALL1 has to be true. Thus our task is to prove:

$\qquad$ {PRE_LOOP}
$\qquad$ **repeat**
$\qquad\quad$ mm_data_select : = mm_data_select − 1
$\qquad\quad$ **call** MAIN_MEM.PUSH_G

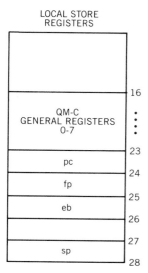

Figure 14.16 Layout of the QM-1 local store.

  until mm_data_select = reg.inst_reg.c
  {INT_CALL1}

Consider the representation of the QM-C pointers, counters, and general registers within the QM-1 local store (Fig. 14.16). Informally, at the end of the jth iteration of the loop ($1 \leq j \leq 25 - \text{r\_lowest}_0$) the stack in locations main_mem [sp_0 .. $sp_0 - j + 1$] will hold the value of the elements in local_store [25 .. $26 - j$], and stack pointer (sp) will point to the next empty element on the stack. Formally, this state of affairs is captured in the following invariant:

INV_CALL:
  $(1 \leq j \leq 25 - \text{r\_lowest}_0)$ (main_mem[$sp_0 - j + 1$ .. sp_0] =
      reg[$26 - j$ .. 25]
& reg.sp : ls_register | mm_addr_so[n_sp] :
      ls_register = $sp_0 - j$
& mm_data_select : **seq**[5..0]**bit** = $26 - j$
& mm_addr_select : **seq**[5..0]**bit** = 28

As usual in such cases, we prove this invariance by induction.

1. For $j = 1$ (i.e., at the end of the first iteration), we are required to show that:

  INV_CALL1:
    main_mem[sp_0 .. sp_0] = reg[25 .. 25]

& reg.sp : ls_register|mm_addr_so[n_sp] :
$$\text{ls\_register} = sp_0 - 1$$
& mm_data_select : **seq**[5..0]**bit** = 25
& mm_addr_select : **seq**[5..0]**bit** = 28

When the loop is entered for the first time, PRE_LOOP holds. This leads to the formula:

{reg.sp : ls_register|mm_addr_so[n_sp] : ls_register = b_0
& mm_addr_select : **seq**[5..0]**bit** = 28
& mm_data_select : **seq**[5..0]**bit** = 26}

$$\text{mm\_data\_select} := \text{mm\_data\_select} - 1$$

{reg.sp : ls_register|mm_addr_so[n_sp] :
ls_register = sp_0
& mm_addr_select : **seq**[5..0]**bit** = 28
& mm_data_select : **seq**[5..0]**bit** = 25

which is evidently true. At this stage, the mechanism MAIN_MEM is activated through the call to PUSH_G. According to the proof rule for the **proc** statement (with procedure body S), if we wish to prove {P} **call** p{Q} then we must first prove {P}S{Q}. Thus, we have to show that the following holds:

{reg.sp : ls_register|mm_addr_so[n_sp] : ls_register = sp_0 &
mm_addr_select : **seq**[5..0]**bit** = 28 & mm_data_select :
seq[5..0]**bit** = 25}
BODY OF PUSH_G
{INV_CALL1}

Comparing the preconditions of the procedure PUSH_G with the precondition in the formula above, since r_0, t_0 are (free) dummy variables we may substitute the respective literals 28 and 25 for them. That is, the assertion PRE_PUSH_G may be restated as

PRE_PUSH_G1:
 mm_addr_select : **seq**[5..0]**bit** = 28
 & mm_data_select : **seq**[5..0]**bit** = 25
 & mm_addr_so.reg[28] : ls_register = sp_0
 & mm_data_so.reg[25] : ls_register = fp_0

From the proof outline for PUSH_G, the result of executing this procedure is

main_mem[sp_0] : **seq**[17..0]**bit** = fp_0
& main_output : bus = fp_0
& mm_addr_select : **seq**[5..0]**bit** = 28
& mm_data_select : **seq**[5..0]**bit** = 25
& mm_data_so [25] : ls_register = fp_0
& mm_addr_so.reg [28] : ls_register = $sp_0 - 1$

This implies the assertion INV_CALL1.

2. Now, assume as the induction hypothesis that INV_CALL is true after the $(k - 1)$th iteration; we are to show that this assertion holds after the kth iteration (where $k \leq r\_lowest_0$). The state of the stack and the relevant variables at the beginning of the loop body are shown in Fig. 14.17 and given by the assertion:

INV_CALL $k - 1$:
main_mem[$sp_0 - k + 2 .. sp_0$] = reg[$27 - k .. 25$]
& reg.sp : ls.register | mm_addr_so.reg[n_sp] :
$$ls\_register = sp_0 - (k - 1)$$

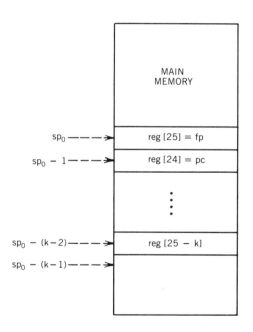

Figure 14.17 State of the stack and other registers before execution of the loop body.

& mm_data_select : **seq**[5..0]**bit** $= 26 - (k - 1)$
& mm_addr_select : **seq**[5..0]**bit** $= 28$

We now wish to show that the following assertion holds after the kth iteration:

INV_CALL k:
  main_mem[$sp_0 - k + 1 .. sp_0$] = reg[$26 - k .. 25$]
& reg.sp : ls_register | mm_addr_so.reg[n_sp] :
    ls_register = $sp_0 - k$
& mm_data_select : **seq**[5..0]**bit** $= 26 - k$
& mm_addr_select : **seq**[5..0]**bit** $= 28$

This leads to the trivially true formula

{INV_CALL $k - 1$}

  mm_data_select := mm_data_select $- 1$

{main_mem[$sp_0 - k + 2 .. sp_0$] = reg[$27 - k .. 25$]
  & reg.sp : ls.register | mm_addr_so[n_sp] :
                ls_register = $sp_0 - (k - 1)$
  & mm_data_select : **seq**[5..0]**bit** $= 26 - k$
  & mm_addr_select : **seq**[5..0]**bit** $= 28$}

On activating PUSH_G we may, as before, substitute the expressions $26 - k$ and 28 for dummy variables r_0, t_0, respectively, so that PRE_PUSH_G becomes:

PRE_PUSH_G $k - 1$:
  mm_addr_select : **seq**[5..0]**bit** $= 28$
& mm_data_select : **seq**[5..0]**bit** $= 26 - k$
    & mm_addr_so.reg[28] : ls_register = $sp_0 - (k - 1)$
    & mm_data_so[$26 - k$] : ls_register = d_0

The result of executing the procedure body, from the proof outline for PUSH_G, is:

main_mem[$sp_0 - (k - 1)$] : **seq**[17..0] **bit** =
            mm_data_so[$26 - k$] | reg[$26 - k$]
& main_output : bus = mm_data_so[$26 - k$]
& mm_addr_select : **seq**[5..0]**bit** $= 28$
& mm_data_select : **seq**[5..0]**bit** $= 26 - k$
& mm_addr_so.reg [28] : ls_register | reg.sp :
            ls_register = $sp_0 - (k - 1) - 1$

In addition, since the condition

$$\text{main\_mem}[sp_0 - k + 2 .. sp_0] = [27 - k .. 25]$$

is unaffected we obtain, by their conjunction:

$\text{main\_mem}[sp_0 - (k - 1) .. sp_0] = \text{reg}[26 - k .. 25]$
$\&\ \text{reg.sp} : \text{ls\_register} | \text{mm\_addr\_so.reg}[28] = sp_0 - k$
$\&\ \quad \text{mm\_data\_select} : \mathbf{seq}[5..0]\mathbf{bit} = 26 - k$
$\&\ \quad \text{mm\_addr\_select} : \mathbf{seq}[5..0]\mathbf{bit} = 28$

which is simply INV_CALL k. This completes the (partial) proof of correctness for the **repeat .. until** statement in Fig. 14.15.

Finally, to show that this statement terminates, note that mem_data_select, starting with an initial value of 26, is decremented by 1 in each iteration and tested against reg.inst_reg.c ($=$ r_lowest$_0$). Thus as long as r_lowest$_0 \leq 25$, the loop is guaranteed to terminate. According to the precondition PRE_CALL, this relation is assumed to hold.

EPILOGUE

My main objective in writing this book was to draw attention to an aspect of computer architecture that has, by and large, been neglected by theorists and practitioners alike: the concept of architecture as a design discipline.

The problem of architecture design seems to involve two major issues. The first of these has to do with what I have called informal design, and the problems it poses are of the following nature: given a set of requirements, how can the architect formulate an efficient search through the design space in order to select an appropriate set of components or subassemblies? What kinds of heuristics can the designer apply in pruning the tree of potential designs? How can we organize the established principles of computer architecture in the form of a knowledge base so that it may aid the design process?

These are difficult and large questions that are common to many spheres of design. In computer architecture in particular, they provide a rich, unexplored territory for further inquiry. In this book I have touched on only two aspects of informal design: the concept of architectural style and the use of empirical data on languages that architectures are intended to support as a basis for exoarchitectural design.

As I have shown in Chapter 13, the latter is a technique that has been adopted by several architects in recent years. Further, the many studies of how programmers use programming languages reveal some program characteristics that are essentially independent of the actual language used. It appears reasonable to promote these characteristics to the status of general principles that should become part of every architect's knowledge base; these can be augmented with other similar principles as further data are accrued.

As regards architectural styles, it seems unlikely that any architect would take serious issue with the assertion that there are recognizable styles in architecture or with the idea that architects, in embarking on a new design, rarely start from scratch. Rather, they draw on their knowledge of or experience with previously developed styles in making key design choices. A more contentious issue might be the thesis advanced in Chapter 12 that a well-articulated library of styles may be used to drive the (informal) design process; and that one may view informal design as an explicit exercise in selecting and combining different styles that may help reduce the search through the design space.

The other major issue concerns the translation of an informal design into a formal description. I have termed this act of translation the formal design process. This monograph has been largely addressed to this issue.

The advantages of formal machine descriptions and the uses to which they may be put are many. Since they have been widely documented both in this book and elsewhere, further elaboration is unnecessary. However, from the perspective of a design discipline, the most attractive feature of a formal description is that it objectifies the design. If the description language is sufficient in its expressive adequacy, as well as in the rigor of its syntax and semantics, one may manipulate, transform, verify, and above all understand the design *completely*.

The formal design process demands a radically different view of computer architecture. We may call it, for want of a better word, a *symbolic* view in which we are two levels removed from the physical machine: a formal description is a symbolic model of the architecture, which in turn is an abstraction of the hardware.

For formal design to succeed as a design technique, two conditions must be met. First, techniques must emerge—partially or fully automated—for translating formal descriptions of architectures into blueprints for hardware realization. In other words, formal architectural design must be integrated into conventional design automation systems. Recent work in this direction has been reported by Hafer and Parker (1982) and Lim (1982). However, much work remains to be done in this domain, in particular in applying such automated systems to the synthesis of real architectures.

The second condition for the success of formal design is an enlargement in the architect's view of his or her subject matter so as to admit this symbolic approach to architecture design.

This is not merely a matter of architects convincing themselves of the scientific merits of formal design. Rather, it demands the removal of certain intellectual, psychological, and even cultural barriers in order to accept a paradigm that is somewhat at odds with the prevailing paradigm of what computer architecture is and what an architect does.

Such paradigm shifts have always been painful, as the history of science has shown; see, for example, Kuhn (1970). But they do sometimes occur. And if we agree that the textbook represents the common corpus of conventional wisdom in a particular branch of knowledge, the signs are encouraging. Among recent textbooks, Baer (1980) devotes an entire chapter to formal description and modeling. Garside (1980) contains a section on ISP, while a slightly earlier publication, Hayes (1978), has an extensive discussion on design methodology.

In the light of this evidence one can only be optimistic that textbooks of the future—which, after all, will provide the educational foundation for the next generation of computer architects—will increasingly place greater emphasis on computer architecture as a design discipline and thereby accelerate the paradigm shift.

APPENDIX A

A DEFINITION OF THE ARCHITECTURAL DESCRIPTION LANGUAGE S*A

INTRODUCTION

The S*A language is designed as a general purpose, high-level procedural language for the description of computer architectures. While such earlier publications as Dasgupta (1981), Dasgupta and Olafsson (1982), and Dasgupta (1982) have documented the rationale underlying the language design, and have provided (mainly through examples) an informal description of its characteristics, a formal definition of S*A has not been published heretofore. This report is intended to fill this gap, and must henceforth be regarded as the official definition of S*A.

In selecting an appropriate form for the language description, I have taken into special account the following points:

1. Since each construct in S*A denotes, in an abstract way, a class of hardware objects or actions, the potential user of S*A must be provided with an explicit description of the correspondence between an S*A construct and the class of hardware entities it represents.

2. A major goal in designing S*A was to provide a vehicle for the verification of architectural designs. A sine qua non for such verification is a definition of S*A in terms of abstract properties that are quite independent of the hardware semantics mentioned in the first point above.

Thus the semantics of S*A is composed of an informal part that establishes the language-hardware correspondence, and a formal part that states the abstract properties of the language.

It must be noted that in spite of the term used and the manner of its presentation, the informal part does not imply inexactness of definition. At all times, the language constructs are described with reference to hardware terms which themselves have precise and well-defined meanings.

In presenting the formal part, I have been greatly influenced by the axiomatic method of programming language definition first proposed by Hoare (1969) and applied to the definition of Pascal by Hoare and Wirth (1973). To my knowledge this method has not so far been applied to the domain of hardware description languages. It consequently posed a novel set of problems in language definition.

I must also record a debt to Brinch Hansen's (1981) paper on Edison, and in particular to his illuminating discussion of the problems of language description.

1. NOTATION, TERMINOLOGY, AND BASIC VOCABULARY

Syntactic entities are enclosed between the angular brackets < and >. The notation {x} denotes zero or more repetitions of the entity x. The notation ↓ y ↑ denotes zero or one instance of the entity y.

In defining the semantics of the language the following notation will be used:

   1. S*A statements will be denoted by $S, S_1, S_2, \ldots$, and so on.
   2. Logical formulas describing properties of data objects will be denoted by P, Q, R.
   3. Assertions will be denoted by H. The most common form of an assertion H will be {P}S{Q}, which states that if P is true before the execution of S then Q is true when S terminates.
   4. Rules of inference are of the form
$$\frac{H_1, H_2, \ldots, H_n}{H}$$
   which states that whenever $H_1, H_2, \ldots, H_n$ are true then H is also true. Occasionally, the alternative notation $H_1, H_2, \ldots, H_n \supset H$ will be used to mean the same thing.
   5. Finally, the symbols *true* and *false* will denote, respectively, predicates that are always true and always false.

1.1. The Basic Vocabulary

<letter> ::= A|B|C|D|E|F|G|H|I|J|K|L|M|N|O|P|Q|R|S|T|U|V
|W|X|Y|Z|
a|b|c|d|e|f|g|h|i|j|k|l|m|n|o|p|q|r|s|t|u|v|
|w|x|y|z|

<digit> ::= 0|1|2|3|4|5|6|7|8|9

<special symbol> ::= +|−|*|/|∧|∨|⌐|⊕|:|;|=>|⌐∧|⌐∨|:=|□|=|
<|>|<=|>=|=|[|]|(|)|.|,|sys|mech|glovar|
privar|seq|bit|array|of|with|value|tuple|stack|
assoc|call|proc|if|do|od|fi|case|endcase|push|
pop|return|endmech|endsys|syn|const|pconst|
bin|oct|dec|hex|type|trap|while|repeat|forever|
sync|await|sig|init|endinit|endproc|chan|when|
act|endtup|else|shl|shr|shlc|shrc|shll|shrl|topval

1.2. Comments

In an S*A text, comments may freely appear enclosed by the symbols /* and */.

2. IDENTIFIERS

Identifiers are symbols used to denote state variables, constants, types, procedures, and functions. All identifiers must be unique.

<identifier> ::= <letter>|<identifier><letter>|<identifier>
<digit>|<identifier>_<identifier>

2.1. Examples

A
inst_buffer_unit1
MEM12

3. INTEGERS

<integer> ::= <bin integer>|<oct integer>|<hex integer>|
<explicit dec integer>|<implicit dec integer>
<bin integer> ::= **bin** <bin digit>{<bin digit>}
<oct integer> ::= **oct** <oct digit>{<oct digit>}
<hex integer> ::= **hex** <hex digit>{<hex digit>}
<explicit dec integer> ::= **dec** <dec digit>{<dec digit>}
<implicit dec integer> ::= <dec digit>{<dec digit>}

```
<bin digit> ::= 0|1
<oct digit> ::= <bin digit>|2|3|4|5|6|7
<dec digit> ::= <digit>
<hex digit> ::= <digit>|A|B|C|D|E|F
```

3.1. Semantics

The syntactic entity <integer> denotes an integer in one of the four number systems binary, octal, hexadecimal, and decimal. A decimal integer can be represented explicitly by prefixing the integer with **dec**.

3.2. Examples

> 86
> **bin** 011010
> **oct** 727
> **hex** 2A4D
> **dec** 512

4. VARIABLE DATA TYPES

```
<type> ::= <simple type>|<structured type 1>|<structured type 2>|
           <type identifier>
<simple type> ::= bit
<structured type 1> ::= <seq>|<tuple>
<structured type 2> ::= <array>|<assoc array>|<stack>
<type identifier> ::= <identifier>

<seq> ::= seq <dim> bit
<dim> ::= [<implicit dec integer> .. <implicit dec integer>]|
          [..]|[.]|[]
<array> ::= array <dim> of <structured type 1>{<with qualifier>}|
            array <dim> of <type identifier>{<with qualifier>}

<assoc array> ::= assoc array <dim> of <structured type 1>|
                  assoc array <dim> of <type identifier>
<stack> ::= stack <stack dim> of <structured type 1>
            {<with qualifier>}|stack<stack dim> of
            <type identifier> {<with qualifier>}
<tuple> ::= tuple <identifier>:<type>{;<identifier>:<type>}
            endtup
<with qualifier> ::= with <identifier>|<with qualifier>,
                     <identifier>
<stack dim> ::= [<implicit dec integer>]
```

VARIABLE DATA TYPES   253

4.1. Semantics

A data type in S*A conforms to the concept of type as suggested by Hoare (1972), that is, (1) a data type determines a particular class of values that may be assumed by a variable or expression of that type; (2) the type of a value denoted by any constant, variable, or expression may be deduced from its form or context without any knowledge of its value as computed at activation time; and (3) associated with each type is a set of one or more primitive operations that can be applied to these values.

S*A contains the simple data type **bit**, and the structured data types sequence, array, assoc array, stack, and tuple.

4.1.1 The data type **bit** consists of the values $\{0,1\}$. Operations defined on type **bit** consist of the logical operators $\{\wedge, \vee, \neg, \oplus, \neg\wedge, \neg\vee\}$ and the arithmetic operators $\{+, -, *, /\}$.
  - **a** Given x, y of type **bit**, x/y is undefined for $y = 0$.
  - **b** Given x of type **bit**, x represents a binary storage element.

4.1.2 Let T denote a sequence type: $\text{seq}[i_h .. i_\ell]$ **bit**
  - **a** $i_h \geq i_\ell$, $i_h, i_\ell \geq 0$
  - **b** Let $i_h \geq i \geq i_\ell$

    Then $T : \{i_h,...,i_\ell\} \rightarrow$ **bit**. That is, $T(i)$ is an element of type **bit**.
  - **c** Let $<bb .. b>_{i_\ell}^{i_h}$ denote a binary string of length $i_h - i_\ell + 1$. Then $T = \{<00 .. 0>_{i_\ell}^{i_h}, <00 .. 1>_{i_\ell}^{i_h}, \ldots, <11 .. 1>_{i_\ell}^{i_h}\}$.

    That is, a **seq** type is defined both as a mapping function and as a set of values.
  - **d** The logical operations $\{\wedge, \vee, \neg, \oplus, \neg\wedge, \neg\vee\}$ are defined on T according to the following rule. Let θ denote a logical operation, and x, y be of type seq $[i_h .. i_\ell]$ **bit** such that

    $x = <x_{i_h}x_{i_h-1} \ldots x_{i_\ell}>$, and $y = <y_{i_h}y_{i_h-1} \ldots y_{i_\ell}>$.

    Then $x \theta y = x_{i_k} \theta y_{i_k}$ for all $i_k : i_h \geq i_k \geq i_\ell$.
  - **e** The arithmetic operations $+, -, *, /$ are defined on T as follows: let $+', -', *', /'$ denote ordinary arithmetic addition, subtraction, multiplication, and integer quotient division. Let x, y be values of type T and **max** be the maximum value ($<11 .. 1>_{i_\ell}^{i_h}$) in the set of values of type T. Then $+, -, *, /$ on x, y are defined by

    (i) $x +' y \leq \mathbf{max} \supset x + y = x +' y$
    (ii) $x \geq y \supset x - y = x -' y$
    (iii) $x *' y \leq \mathbf{max} \supset x * y = x *' y$
    (iv) $y \neq <00 .. 0>_{i_\ell}^{i_h} \supset x/y = x/'y$
  - **f** The shift operations $\{\mathbf{shl}, \mathbf{shr}, \mathbf{shlc}, \mathbf{shrc}, \mathbf{shll}, \mathbf{shrl}\}$ are defined on T as follows. Let:

    $x = <b_i b_{i-1} \ldots b_0>_{i_\ell}^{i_h}$, $b_i, b_{i-1}, \ldots, b_0 \in \{0,1\}$.

Then:

$\text{shl}(x) = <b_{i-1}b_{i-2}..b_00>_{i_\ell}^{i_h}$ (left shift)

$\text{shr}(x) = <0b_ib_{i-1}..b_1>_{i_\ell}^{i_h}$ (right shift)

$\text{shlc}(x) = <b_{i-1}b_{i-2}..b_0b_i>_{i_\ell}^{i_h}$ (left shift circular)

$\text{shrc}(x) = <b_0b_i..b_2b_1>_{i_\ell}^{i_h}$ (right shift circular)

$\text{shll}(x) = <b_{i-1}b_{i-2}..b_01>_{i_\ell}^{i_h}$ (left shift and insert one)

$\text{shrl}(x) = <1b_i..b_2b_1>_{i_\ell}^{i_h}$ (right shift and insert one)

   g The arithmetic, logical, and shift operations defined on type T represent primitive (hardware-defined) functional logic units.

   h In the following paragraphs, let x be of type T. Then x represents any device capable of storing binary strings in the range of values defined by T (see 4.1.2.c) such that elements of the binary string can be accessed in parallel.

   i If the range of T is denoted by "[..]" this signifies that i_h, i_ℓ are unspecified but will be known at design time.

   j If the range of T is denoted by "[]" this signifies that i_h, i_ℓ are unspecified and will not be known at design time, but are defined whenever x is referenced during execution of an S*A description.

   k If the range of T is given as "[.]" this signifies that the range of the sequence may vary dynamically, but with a maximum limit that will be known at design time.

   l Examples:
      (a) **seq** [15 .. 0] **bit**
      (b) **seq** [..] **bit**
      (c) **seq** [.] **bit**

4.1.3 Array type : **type** T = **array** [i_ℓ .. i_h] **of** T_0

   a $i_\ell < i_h$, i_ℓ, $i_h \geq 0$

Let $i \in \{i_\ell,..,i_h\}$

   b T : $\{i_\ell,..,i_h\} \rightarrow T_0$; that is, T(i) is an element of type T_0.

   c Given x of type T, the value of x, denoted v(x), is the ordered k-tuple of values of type T_0, where $k = i_h - i_\ell + 1$.

   d Let x be of type T. Then x represents any device capable of storing an ordered set of values each of type T_0 such that one and only one element of x can be accessed at a time.

   e Examples:
      (a) **array** [0 .. 31] **of seq** [15 .. 0] **bit**
      (b) **array** [0 .. 15] **of** register

   f Examples:
register arrays, main memory, control store can be represented by the type **array**

Let x be of type T.

   g If the range of an array type T is denoted by "[..]" this signifies that i_h, i_ℓ are unspecified but will be known at design time.

VARIABLE DATA TYPES

h If the range is denoted by "[]" this signifies that i_ℓ, i_h will not be known at design time but will be defined whenever x is referenced during execution of an S*A description.

i If the range of T is given as "[.]" this signifies that the range of the array may vary dynamically but with a maximum size that is known at design time.

4.1.4 Array-with-pointer type:

type T = **array** $[i_\ell \ .. \ i_h]$ **of** T_0 **with** $v_1, v_2, \ldots, v_n$

a $i_\ell < i_h$

b $v_1, v_2, \ldots, v_n$ must be (declared elsewhere as) of type **seq**.

Let $i \in i_\ell \ .. \ i_h$. Let $v_j \in \{v_1, \ldots, v_n\}$ be of type T_j. Let **intval** (v_j) denote possible (decimal) integer values of v_j, $1 \leq j \leq n$.

c $i_\ell \leq$ **intval** $(v_j) \leq i_h$.

d $T : T_j \rightarrow T_0$. That is, $T(v_j)$ is an element of type T_0.

e Let $\min_j \ .. \ \max_j$ denote the range of **intval** $(v_j)(1 \leq j \leq n)$. Then the elements of T are $T(\min_j)$, $T(\min_j + 1)$, $\ldots$, $T(\max_j)$ with respect to $v_j (1 \leq j \leq n)$.

f Let x be of type T. Then x represents any device capable of storing an ordered set of values each of type T_0 such that one and only one element of x can be accessed at a time. Further, elements of x can only be accessed by using $v_j \in \{v_1, \ldots, v_n\}$ as a "pointer" variable.

g Same as 4.1.3.g.

h Same as 4.1.3.h.

i Same as 4.1.3.i.

j Examples:

    (a) **array** [0 .. 8191] **of word with** mar

    (b) **array** [0 .. 1023] **of seq** [15 .. 0] **bit**
         with mir1, mir2, mir3.

4.1.5 Tuple type: **type** T = **tuple** $S_1 : T_1; \ldots ; S_n : T_n$ **endtup**

a Type T is the Cartesian product of $T_1, \ldots, T_n$; that is, $T = T_1 \times T_2 \times \ldots \times T_n$.

b Let x be of type T. Then x is a structured storage device consisting of a number of components or fields $S_1, \ldots, S_n$ of types $T_1, \ldots, T_n$ respectively. A field S_i can be referenced by using either of the notations $x.S_i$ or S_i. The latter is valid, however, only if S_i is a system-wide unique name.

c There are no valid operations defined on T. Valid operations on the field types $T_1, \ldots, T_n$ are determined by the nature of the types $T_1, \ldots, T_n$ themselves.

d Example:
        tuple opcode : **seq** [3 .. 0] **bit**;
              addr   : **seq** [11 .. 0] **bit**;
             index : **seq** [1 .. 0] **bit**
    endtup

4.1.6 Stack type:

    type T = **stack** [i] **of** T_0
    type T = **stack** [i] **of** T_0 **with** $v_1, \ldots, v_n$ ($n \geq 1$)

a $v_1, v_2, \ldots, v_n$ are of type **seq**.
b **intval** $(v_j) \leq i$ $(1 \leq j \leq n)$.
c T_0 is of type **seq** or **tuple**.
d Let v_j be of type T_j. Then $T : T_j \to T_0$.
e A **stack** data type is an ordered collection of storage elements, all of the same type (either **seq** or **tuple**) such that only its "top" element is accessible. In the first form of a **stack** type (without the **with** clause) any variable v_j satisfying 4.1.6.a,b can serve as a stack pointer (i.e., define the current top element). In the second form, the elements $v_1, \ldots, v_n$ are names of variables of type **seq** that serve as stack pointers. The integer i denotes the maximum stack depth.
f The standard procedures **push** and **pop** are defined on T as follows: let x be of type T and x_0 be of type T_0. Let v_j be a stack pointer (in the second form $v_j \in \{v_1, \ldots, v_n\}$). Let **length** (x) be the number of elements of type T_0 in x. Then

    (a) $\{$**intval** $(v_j) = v_j^0 \wedge$ **length** $(x) = \ell_0 \wedge \ell_0 < i \wedge x_0 = x_0^0\}$ **push** $(x[v_j], x_0)$
          $\{x[v_j] = x_0 = x_0^0 \wedge$ **intval** $(v_j) = v_j^0 + 1 \wedge$ **length** $(x) = \ell_0 + 1\}$

    (b) $\{$**intval** $(v_j) = v_j^0 \wedge$ **length** $(x) = \ell_0 \wedge \ell_0 \geq 1 \wedge x[v_j] = x^0\}$
          pop $(x[v_j], x_0)$
          $\{x_0 = x^0 \wedge$ **intval** $(v_j) = v_j^0 - 1 \wedge$ **length** $(x) = \ell_0 - 1\}$

g The standard function **topval** is defined by **topval** $(x) = x[v_j^0]$ $\ni$ **intval** $(v_j) = v_j^0$.

4.1.7 Associative array type: **type** T = **assoc array** $[i_\ell \ .. \ i_h]$ **of** T_0.

a A variable x of type **assoc array** is an array of elements of type T_0 such that the elements of x can only be accessed associatively. That is, given a value y, a word of x is selected for reading from or writing into, by comparing y simultaneously with a specified subpart ("key") of every element in x. The data type **assoc array** is thus used to represent associative memories.
b $i_\ell < i_h$.
c T_0 is of type **seq** or **tuple**.
d There are no primitive operations or standard procedures defined on **assoc array**.
e Example:

    assoc array [0 .. 511] **of**
        tuple segment_id : **seq**[15 .. 0] **bit**;
               base_addr : **seq**[21 .. 0] **bit**;

length : seq[9 .. 0] **bit**
endtup

5. CONSTANT TYPES

<const type> : : = <bin const type>|<oct const type>|
       <dec const type>|<hex const type>
<bin const type> : : = **bin**(<const size>)<bin digit>{<bin digit>}
<oct const type> : : = **oct**(<const size>)<oct digit>{<oct digit>}
<dec const type> : : = **dec**(<const size>) ↓ <sign> ↑ <dec digit>
       {<dec digit>}
<hex const type> : : = **hex**(const size) <hex digit>{<hexdigit>}
<const size> : : = <implicit dec integer>
<sign> : : = + | −

5.1. Semantics

A constant (data) type denotes a specific (but potentially infinite) class of storage devices whose size in bits is given by <const size> and which has the *invariant* value denoted by the syntactic clause following <const size>.

Constant types are used to represent read only storage devices of fixed (wired-in) values. Let x be an instance of a specific constant type. Then x is termed a *constant object*.

Consider the constant type T : <int type>(i)j, where <int type> ∈ {**bin, oct, dec, hex**}. Let x be of type T, and let binval (x), octval (x), decval (x), and hexval (x) denote the value of x in the binary, octal, decimal, and hexadecimal number systems. Then:

5.1.1 <int_type> = **bin** ⊃ binval (x) = j
   <int_type> = **oct** ⊃ octval (x) = j
   <int_type> = **hex** ⊃ hexval (x) = j
   <int_type> = **dec** ⊃ decval (x) = j

6. DECLARATIONS OF VARIABLES AND CONSTANT OBJECTS

6.1. Variables

<var> : : = <simple_var>|<var>.<simple_var>
<simple_var> : : = <identifier>|<identifier>[<subscripts>]|
       <identifier>[<subscripts>][<subscripts>]
<subscripts> : : = <arith expr>|<arith expr>..<arith expr>

6.2. Type Declarations

<type decln> : := **type** <identifier>{,<identifier>} = <type>
<type decln set> : := <type decln>|<type decln set>;<type decln>

6.3. Variable Declaration

<glovar decln> : := **glovar** <var decln>
<privar decln> : := **privar** <var decln>
<var decln> : := <identifier>{,<identifier>} : <type>{: <type>}

6.4. Synchronizer Declaration

<sync decln> : := **sync** <sync list>
<sync list> : := <identifier>{,<identifier>} : <sync type>
             {(<init val>)}
<sync type> : := **bit**|<seq>
<init val> : := <bin digit>|<implicit dec integer>

6.5. Channel Declaration

<chan decln> : := **chan** <chan list>;
<chan list> : := <identifier>{,<identifier>} : <chan type>
<chan type> : := **bit**|<seq>

6.6. Constant Declaration

<const decln> : := **const**<const list>|**pconst**<const list>
<const list> : := <identifier>{,<identifier>} : <const type>

6.7. Synonym Declaration

<syn decln> : := **syn** <synonym>{,<synonym>}
<synonym> : := <identifier>{:<type>} = <var>

6.8. Semantics

In general, the purpose of a declaration is to introduce a named object (e.g., type, variable, procedure) and to prescribe its properties. This section is concerned only with data object declarations, while later sections deal with the declarations of "active" objects such as procedures and mechanisms.

Data objects are broadly of two types: *variables* (whose values may change over time) and *constants* (whose values remain invariant over time). In an S*A description the declaration of a variable or constant data object of a given type T denotes the existence, in the architecture being designed and/

DECLARATIONS OF VARIABLES AND CONSTANT OBJECTS   259

or described, of a storage device whose abstract properties are prescribed by the properties of the data type T.

The declaration of a pseudoconstant (see 6.6 and 6.8.4) of a given constant type T' denotes simply a single value that satisfies the properties of type T'. The architecture being described does not, however, contain an actual storage device corresponding to the pseudoconstant. The pseudoconstant thus denotes a pseudoobject or storage device.

Each distinct data object thus declared is a specification of a distinct (real or pseudo) storage device. That is, given two declarations (neither of which is a synonym declaration) that define storage devices S_1, S_2, S_1 is not contained in S_2 ($S_1 \not\subseteq S_2$), nor is S_2 contained in S_1 ($S_2 \not\subseteq S_1$).

6.8.1 A type declaration

$$\text{type } T = T'$$

where T' specifies a **bit, seq, array, stack,** or **assoc array** type, introduces a class of possible objects of type T that will satisfy the properties of type T' as stated in Section 4.

6.8.2 Variables are always declared as global or private with respect to a mechanism or a system (see Sections 11 and 12). Variable declarations

$$\textbf{glovar } x : T$$
$$\textbf{privar } y : T'$$

introduce named data objects of types T, T' respectively. Variables x and y thus have the properties associated with types T, T' as stated in Section 4.

6.8.3 A constant declaration

$$\textbf{const } x : T$$

where T is a constant type introduces a named (real) storage device of type T that has the properties associated with type T as stated in Section 5.

6.8.4 A pseudoconstant declaration

$$\textbf{pconst } x : T$$

where T is a constant type introduces a named (pseudo) storage device of type T that has the properties associated with type T as defined in Section 5.

6.8.5 Examples:

    type register, word = **seq** [15 .. 0] **bit**
    glovar memory : **array**[0 .. 4095] **of** word
    privar local_store : **array**[0 .. 31] **of** register
    glovar ctl_stack : **stack**[8] **of seq**[15 .. 0] **bit**
    const one : **dec** (16) 1
    pconst indir : **dec** (4) 8
    privar inst_reg : **tuple**

opcode : **seq**[12 .. 0] **bit**;
operand : **seq** [10 .. 0] **bit**;
indx : **bit**
endtup

6.8.6 It is possible to declare a variable as an instance of two or more data types. Such variables are termed *multitype* variables, and the corresponding declarations are multitype declarations. Given a multitype declaration, for example:

$$\textbf{glovar } x : T_1$$
$$: T_2$$
$$\ldots$$
$$: T_n$$

x is defined as simultaneously satisfying the properties associated with types $T_1, \ldots, T_n$.

a Let $|x|^{T_i}$ denote the length of the storage device x in bits, in its incarnation as an instance of type T_i. Then a multitype declaration

$$\textbf{glovar } x : T_1 : T_2 \ldots : T_n$$

or

$$\textbf{privar } x : T_1 : T_2 \ldots : T_n$$

is invalid if $|x|^{T_i} \neq |x|^{T_j}$ for any pair i, j, $(1 \leq i \leq n, 1 \leq j \leq n, i \neq j)$

6.8.7 Example:

glovar inst_reg : **seq**[23 .. 0] **bit**
: **tuple**
opcode : **seq**[4 .. 0] **bit**;
opd   : **seq**[14 .. 0] **bit**;
indir   : **bit**;
regno : **seq**[2 .. 0] **bit**
endtup

6.8.8 A synchronizer declaration:

sync x : T

where T is of type **bit** or of type **seq** introduces a named variable of type T with the following properties:

a The standard procedures **await** x and **sync** x are defined on a synchronizer according to the following semantic rules:

(a) $\{x = x_0 > 0\}$ **await** x $\{x = x_0 - 1 \geq 0\}$
$\{x = x_0 = 0\}$ **await** x $\{false\}$
(b) $\{x = x_0 : 0 \leq x_0 < MAX\}$ **sig** x$\{x = x_0 + 1 > 0\}$

where MAX is the maximum integer valued state possible for x. Intuitively, an **await** operation will never terminate as long as the value of x is 0; when $x > 0$, the **await** decrements the value of x and terminates. The **sig** operation is only defined if x < MAX. In that case it simply increments x.

DECLARATIONS OF VARIABLES AND CONSTANT OBJECTS 261

    b No other operations are defined on x.

6.8.9 A channel declaration:

$$\textbf{chan } x : T$$

where T is of type **bit** or **seq** denotes a named device with the following properties:

    a If x is of type **seq** it satisfies properties 4.1.2.a–4.1.2.c.
    b Arithmetic, logical, and shift operations are *not* defined on channels.

Let x, x' be channels of type **bit** and **seq**$[i_h .. i_\ell]$ **bit** respectively. Let $i \in \{i_h .. i_\ell\}$. Then:

    c Possible testable values of x are 0 (x = 0), 1 (x = 1), the 0-to-1 transition (0 < x < 1), and the 1-to-0 transition (1 > x > 0).
    d Possible testable values of x'[i] are 0, 1, 0-to-1 transition (0 < x'[i] < 1), and 1-to-0 transition (1 > x'[i] > 0).
    e Intuitively, channels are representations of communication paths. Channels resemble ordinary **bit** or **seq** type variables in that they can be assigned binary values. Their binary state can also be sampled. In addition, state transitions in channel bits (from 0 to 1 or vice versa) are testable conditions. This comprises the basic abstract distinction between channels and other variable data objects.

6.8.10 The simplest form of a synonym declaration:

$$\textbf{syn } x = y$$

where y is the name of another object, associates a new name x with the object designated by y such that:

    a Within the scope of the synonym declaration (see Sections 11 and 12), for every occurrence of y in an executional statement, x may be substituted, and vice versa.
    b If y identifies a simple variable or has the subscripted form $x[c_1 .. c_2]$ (where c_1, c_2 are integer constants) then the synonym is said to be *statically determined* (or simply, *static*).

Let P be an assertion and $P^{v_1}_{v_2}$ be the same assertion with all free occurrences of v_1 being replaced by v_2. Then for static synonyms

    c $P \supset P^x_y \,;\, P \supset P^y_x$
    d If y is a subscripted variable of the form $x[v_1 .. v_2]$ or $x[v]$ (where v_1, v_2, v are themselves variable identifiers) then the synonym is said to be *dynamically determined* (or simply, *dynamic*). Dynamic synonyms satisfy, in addition to 6.8.10.a, the following properties:
    e An occurrence of x in an executional statement, at a time instant t of the statement's execution, associates the name x with the object designated by $z[v_1 .. v_2]$ (or $z[v]$) at that instant t.
    f $P \supset P^x_y; P \supset P^y_x$ at the instant t.

g Examples:
Given a variable x declared as

glovar x : **array**[0 .. 63] **of seq**[7 .. 0] **bit**

the following synonyms may be declared:
(a) (static) **syn** y = x[0 .. 7]
(b) (dynamic) **syn** z = x[v_1 .. v_2]

where v_1, v_2 are of type **seq**.

6.8.11 The more general form of a synonym declaration is

$$\text{syn } x : T_1$$
$$: T_2$$
$$\ldots$$
$$: T_n = y$$

where x is the new name, $T_1, \ldots, T_n$ are associated alternative data types, and y is the identifier of the referenced data object.

a Same as 6.8.10.a.
b Depending on whether the synonym is static or dynamic, property 6.8.10.c or properties 6.8.10.e,f will hold respectively.
c Within the scope of the synonym declaration, the object designated by x and y simultaneously satisfies the properties associated with types $T_1, \ldots, T_n$, as well as the type T associated with x.
d Let $|x|^{T_i}$ denote the length in bits of the storage device referenced by x in its incarnation as an instance of T_i, and $|y|^T$ the length in bits of the device referenced by y. Then a synonym declaration of this form is invalid if $|x|^{T_i} \neq |x|^{T_j}$ for any pair i, j ($1 \leq i \leq n$, $1 \leq j \leq n$, $i \neq j$) or if $|x|^{T_i} \neq |y|^T$ for $1 \leq i \leq n$.
e Example:

```
type inst_format1 = tuple opcode : seq[3 .. 0] bit;
                          opd1 : seq[5 .. 0] bit;
                          opd2 : seq[5 .. 0] bit
                    endtup;
type inst_format2 = tuple opcode : seq[6 .. 0] bit;
                          reg : seq[2 .. 0] bit;
                          opd2 : seq[5 .. 0] bit
                    endtup;
glovar inst_reg : seq[15 .. 0] bit;
    . . . . .
syn inst : inst_format1
         : inst_format2 = inst_reg;
```

7. EXPRESSIONS

<arop> ::= + | − | * | /
<unarop> ::= −

```
<binlogop> ::= ∧|∨|⊕|⌐∧|⌐∨
<unlogop> ::= ⌐
<logop> ::= <binlogop>|<unlogop>
<shiftop> ::= shl|shr|shlc|shrc|shll|shrl
<arith expr> ::= <arith term>|<unarop><arith term>
<arith term> ::= <arith factor>|<arith factor><arop><arith factor>
<arith factor> ::= <var>|<integer>|(<arith expr>)|<arith expr>
<bool expr> ::= <bool term>|<unlogop><bool term>|
                <shiftop><bool factor>
<bool term> ::= <bool factor>|<bool factor><binlogop>
                <bool factor>
<bool factor> ::= <var>|<implicit bin integer>|(<bool expr>)
<funct designator> ::= <proc reference>
<expr> ::= <arith expr>|<bool expr>|<funct designator>
<rel expr> ::= <expr><rel op><expr>
<rel op> ::= <|<=|>|>=|=|⌐=
```

7.1. Semantics

An expression consists of variables, constants, operators, and functions and specifies a rule for computing a value. The usual rules of operator precedence apply in the case of arithmetic expressions. Thus, the unary " − " has the highest precedence, followed by the binary operators "*" and "/." The operators " + " and " − " have the lowest precedence.

In Boolean expressions the "⌐" and the shift operators have the highest precedence. All the other logical operators have equal precedence. Relational operators have the lowest precedence.

Sequences of operators of the same precedence are executed in left-to-right order; however, parentheses may be used in the usual manner to alter the operator precedence.

In S*A, evaluation of an expression (other than a function designator) denotes the activation of a combinatorial circuit that accepts as inputs the values in the specified objects (variables and constants) and produces an output according to the operators specified in the expression.

A function designator denotes a function (see Section 10) that, when activated, returns a value.

The value of a nonrelational expression is always of type **bit** or **seq**, since arithmetic, Boolean, and shift operators can only apply to these types (see Section 4). The possible values of a relational expression are 1 and 0, representing the logical values "true" and "false."

7.1.1 Examples:

```
shl local_store[8]
x ∧ y
gpr[6] + inst_reg.opcode + carry_in
stack_top_value
inst_reg.indirect = 0
```

8. STATEMENTS

<stmt> : := <unlabeled stmt> | <label> : <unlabeled stmt>
<unlabeled stmt> : := <simple stmt> | <structured stmt>
<label> : := <identifier>
<simple stmt> : := <assignment stmt> | <proc stmt> | <trap stmt> |
     <goto stmt> | <sync stmt> | <exit stmt> |
     <retn stmt>
<assignment stmt> : := <var>{,<var>} := <expr>
<proc stmt> : := **call** <proc reference> | **act**<proc reference>
<trap stmt> : := **trap** <proc reference>
<goto stmt> : := **goto** <label>
<sync stmt> : := **await**<var> | **sig**<var> | **await**<predicate>
<predicate> : := <bool expr> | <rel expr>
<exit stmt> : := **exit**
<retn stmt> : := **return**
<proc reference> : := <mech identifier>.<identifier>
<mech identifier> : := <identifier>
<structured stmt> : := <compound stmt> | <cond stmt> |
     <repetitive stmt>
<compound stmt> : := <stmt>{;<stmt>} | <stmt>{□<stmt>} |
     do <compound stmt> **od**
<cond stmt> : := <await stmt> | <if stmt> | <case stmt>
<await stmt> : := **await** <var> **do** <stmt> **od** |
     await <predicate> **do** <stmt> **od**
<if stmt> : := **if** <predicate> => <stmt>{ ‖ <predicate> =>
     <stmt>} ↓ ‖ **else** => <stmt> ↓ **fi**
<case stmt> : := **case** <var> **of** <caselist el>{,<caselist el>}
     endcase
<caselist el> : := <case label list> : <stmt>
<case label list> : := <case label>{,<case label>}
<case label> : := <implicit dec integer>
<repetitive stmt> : := <while stmt> | <repeat stmt> | <forever stmt>
<while stmt> : := **while** <predicate> **do** <stmt> **od**
<repeat stmt> : := **repeat** <stmt> **until** <predicate>
<forever stmt> : := **forever do** <stmt> **od**

8.1. Semantics

In S*A, a statement specifies one or more actions to be performed by the hardware. A statement is said to be *executed* when the corresponding actions are performed by circuitry through issuance of appropriate signals from some control unit.

 Statements are basically simple or structured. Simple statements signify indivisible units of action and their meanings are defined formally by axioms.

Structured statements are composed of (more elementary) statements and their meanings are formally specified in terms of rules of inference that allow the actions of structured statements to be deduced from those of their constituents.

8.2. Simple Statements

8.2.1 The assignment statement serves to replace the current value of a variable by the value of the expression. Formally:
$$\{P_E^x\} \; x := E\{P\}$$
The general form of the assignment statement is
$$x_1, x_2, \ldots, x_n := E$$
Informally, the meaning of this statement is: evaluate E to obtain a value v. Then assign v simultaneously to $x_1, \ldots, x_n$. More formally, let $P_E^{x_1,\ldots,x_n}$ denote the predicate P with all free occurrences of $x_1, \ldots, x_n$ replaced by E. Then the multiple assignment statement satisfies the rule
$$\{P_E^{x_1,\ldots,x_n}\} \; x_1, \ldots, x_n := E \; \{P\}$$

8.2.2 The procedure call statement **call p** initiates the named procedure p and transfers control to p. The **call** statement terminates only when control is returned from p. Formally:
$$\{P\} \; p \; \{Q\} \supset \{P\} \; \textbf{call} \; p \; \{Q\}$$
provided that every variable in P, Q is defined within the environment of the **call** statement (see Section 11 for an explanation of environment).

8.2.3 The activate statement **act p** initiates the named procedure p and terminates. The statement following the activate statement then begins execution in parallel with p. The procedure p begins execution only on termination of **act** p. Formally:
$$\{Q\} \; \textbf{act} \; p \; \{Q\}$$

8.2.4 The **trap p** statement is identical in meaning to **call p**, except that control may not return from p to the procedure containing the **trap**. Thus, the statement **trap p** may never terminate. Its semantics is given formally by the rule:
$$\{P\} \; p \; \{Q\} \supset \{P\} \; \textbf{trap} \; p \; \{Q\}$$
if and only if (1) every variable in P, Q is defined within the **trap** statement's environment; and (2) the **trap** statement terminates.

8.2.5 The semantics of the **goto** statement **goto L**, is defined according to the proof rule proposed by Alagic and Arbib (1978). Let S be a statement that contains one or more **goto L** statements, all for the same label L, but does *not* contain "L:" as a label. Then S is termed an *L-statement*.

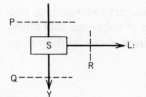

Figure A.1

The formula {P} S {Q} {L : R} for an L-statement means that if the computation state satisfies P on starting S, then on normal exit from S the state satisfies Q, while on an exit from S via a **goto** L, the state satisfies R (Fig. A.1). Thus, for the **goto** statement itself, the rule is:

$$\{P\} \text{ goto } L \text{ \{false\} } \{L : P\}$$

8.2.6 The basic synchronization facilities in S*A are provided by means of the standard procedures **await** and **sig** defined on synchronizing variables according to 6.8.8. Additionally, S*A includes an **await** B statement where B is a predicate. This statement terminates only when B is true. Formally:

$$\{P\} \text{ await } B \text{ } \{P \wedge B\}$$

8.2.7 The **exit** statement causes the procedure p containing the statement to terminate. That is, control is explicitly forced to the end of the procedure. The **exit** is thus effectively a **goto** statement that transfers control to the end of the procedure.

Let the procedure p be denoted by **proc** p; . . . **endproc** and suppose L: is an imaginary label prefixing the **endproc** symbol. Then the **exit** statement can be considered an L-statement (Fig. A.2; cf. also 8.2.5). The corresponding proof rule is, therefore:

$$\{P\} \text{ exit } \{false\} \text{ } \{L : P\}$$

8.2.8 The **return** statement within a procedure p causes control to return to the procedure that originally invoked p. This action is only valid if p had been invoked by **call** or **trap** statements (but not by an **act** statement). The proof rule for **return** statements is identical to that for **exit** statements. Let L: be an imaginary label prefixing the **endproc** symbol for procedure p. Then:

$$\{P\} \text{ return } \{false\} \text{ } \{L : P\}$$

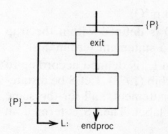

Figure A.2

STATEMENTS

Note. The **return** statement is a means of explicitly forcing a return of control from the invoked to the invoking procedure. In its absence, a procedure p that was the object of a **call** or **trap** statement would automatically return control to the parent procedure when control reached the end of p. However if p was the object of an **act** statement, then on reaching the end, p would simply stop.

8.3. Structured Statements

In general, there are three classes of structured statements. Of these, *compound* statements allow the sequential or parallel composition of simpler statements, *conditionals* allow actions to be taken or to be selected based on specified tests of the machine state, and *repetition* statements permit iterative execution of other statements.

8.3.1 The *sequential* statement $S_1; S_2$ specifies that S_1 and S_2 are to be executed in the order they are written. Its proof rule is:

$$\frac{\{P\}\ S_1\ \{Q\}\ ,\ \{Q\}\ S_2\ \{R\}}{\{P\}\ S_1;\ S_2\ \{R\}}$$

8.3.2 The *parallel* statement $S_1 \square S_2$ specifies the simultaneous execution of S_1 and S_2. The next statement in sequence begins execution only when both S_1 and S_2 have terminated. Formally, provided S_1, S_2 are dynamically disjoint[1]

$$\frac{\{P_1\}\ S_1\ \{Q_1\}\ ,\ \{P_2\}\ S_2\ \{Q_2\}}{\{P_1 \wedge P_2\}\ S_1 \square S_2\ \{Q_1 \wedge Q_2\}}$$

Note. The symbols **do, od** may be used as left and right delimiters of both sequential and parallel statements.

8.3.3 **Await** statements are of two forms: **await** x **do** S **od,** where x is a synchronizer, and **await** B **do** S **od,** where B is a predicate. In the first form the statement S will execute if and only if the synchronizer $x \geq 1$. Thus, **await** x **do** S **od** is semantically equivalent to the statement sequence **await** x; S, except that the former is an *indivisible* statement. Thus, for example, it is not possible to use a **goto** statement to reach a statement inside S from some other statement outside.

In the second form, the Boolean condition B is continuously evaluated until it is true, at which point S begins execution.

The proof rules for these two statements are:

a $\quad \dfrac{\{Q\}\ S\ \{R\}\ ,\ \{P\}\ \textbf{await}\ x\ \{Q\}}{\{P\}\ \textbf{await}\ x\ \textbf{do}\ S\ \textbf{od}\ \{R\}}$

b $\quad \dfrac{\{P \wedge B\}\ S\ \{Q\}}{\{P\}\ \textbf{await}\ B\ \textbf{do}\ S\ \textbf{od}\ \{Q\}}$

---

[1] Two statements S_1, S_2 are said to be *dynamically disjoint* if during the period that S_1, S_2 are both in execution, their data resource sets are disjoint.

8.3.4 The **if** statement

$$\textbf{if } B_1 => S_1 \parallel B_2 => S_2 \parallel \ldots \parallel B_n => S_n \textbf{ fi}$$

where $B_1, \ldots, B_n$ are predicates, is strictly deterministic in that at most one predicate B_i can be true; the corresponding statement S_i is selected for execution. The clauses $B_i => S_i$ are thus order-independent. The **if** statement satisfies the rule:

$$\frac{\{P \wedge B_i\} S_i \{R\} (\forall_i : 1 \leq i \leq n) , \{P \wedge \neg (\bigvee_{i=1}^{n} B_i) \supset R\}}{\{P\} \textbf{ if } B_1 => S_1 \parallel \ldots \parallel B_n => S_n \textbf{ fi } \{R\}}$$

For the sake of conciseness, the statement form "**if** $B_1 => S_1 \parallel B_2 => S_2 \parallel \ldots \parallel B_{n-1} => S_{n-1} \parallel \neg(B_1 \vee B_2 \vee \ldots \vee B_{n-1}) => S_n$ **fi**" can be textually replaced by the statement "**if** $B_1 => S_1 \parallel \ldots \parallel B_{n-1} => S_{n-1} \parallel$ **else** $=> S_n$ **fi**."

8.3.5 The most general form of the **case** statement is:

    case V **of**
        $L_{11}, L_{12}, \ldots, L_{1n_1} : S_1$
        $L_{21}, L_{22}, \ldots, L_{2n_2} : S_2$
        . . . .
        $L_{m_1}, L_{m_2}, \ldots, L_{mn_m} : S_m$
    endcase

where V is a **seq** type variable; $L_{11}, L_{12}, \ldots, L_{mn_m}$ are decimal valued integers such that V may take on these and only these values; and $S_1, \ldots, S_m$ are statements. If the value of V is in the set $\{L_{k_1}, \ldots, L_{kn_k}\}$ ($1 \leq k \leq m$), the corresponding statement S_k is selected for execution. The **case** statement terminates when the selected statement S_k terminates.

Let V_k signify the value of V in the set $\{L_{k_1}, \ldots, L_{kn_k}\}$. Then

$$\frac{\{P \wedge V = V_k\} S_k \{R\} (\forall_k : 1 \leq k \leq m)}{\{P\} \textbf{ case } V \textbf{ of } \ldots \textbf{ endcase } \{R\}}$$

8.3.6 There are basically two repetition statements, namely, **while B do S od** and **repeat S until B**, where B is a predicate and S a statement. In each iteration through the **while** statement, B is first evaluated and if it is true S is executed. In each iteration through the **repeat** statement, S is first executed and then B is evaluated. If B is true control returns to S. More formally, the semantics are given by:

  a $\dfrac{\{P \wedge B\} S \{P\}}{\{P\} \textbf{ while } B \textbf{ do } S \textbf{ od } \{P \wedge \neg B\}}$

  b $\dfrac{\{P\} S \{Q\} , \{Q \wedge \neg B\} S \{Q\}}{\{P\} \textbf{ repeat } S \textbf{ until } B \{Q \wedge B\}}$

Note. P and Q in 8.3.6.a and 8.3.6.b respectively are *loop invariants*.

PROCEDURE DECLARATIONS

A special case of the repetition statement is the nonterminating repetition, **forever do** S **od**. This is functionally equivalent to the statement **while** true **do** S **od**. The **forever** statement satisfies the axiom:

$$\{P\} \text{ \textbf{forever do} } S \text{ \textbf{od}} \{\text{false}\}$$

8.3.7 Examples:
 (a) **do** inst_reg[17 .. 11] := 0
   □ inst_reg[10 .. 0] := operand_buffer
  od
 (b) pc := pc + 1;
   act main_mem.read □ **act** inst_dec.select
 (c) **repeat**
    mm_data_select := mm_data_select − 1;
    call main_mem.push
   until mm_data_select = inst_reg.c
 (d) **if** i < 12 => active [i] := 1 □ allow_nano := 1
   ‖ **else** => active [i] := 1 □ allow_micro := 1
   fi
 (e) **forever do**
    await ibuffer_empty
     do await packet_sent **do** input_buffer := bus.a **od**;
      sig ibuffer_full
     od
   od

9. PROCEDURE DECLARATIONS

<proc> ::= <public proc>|<private proc>
<public proc> ::= **proc** <identifier>;{<syn decln>}
        <stmt> **endproc**
<private proc> ::= **proc** <identifier> **priv**; {syn decln>}
        <stmt> **endproc**

9.1. Semantics

9.1.1 A procedure in S*A is a specification of the dataflow necessary for realizing a particular architectural function. The specified dataflow implies an underlying data path structure in the physical computer that is not defined explicitly in an S*A description.

9.1.2 Procedures are either private or public relative to the mechanism in which they are defined (cf. Section 11). A public procedure may be activated by initiation statements (cf. Section 12) or by public procedures inside other mechanisms, by means of **call** and **act** statements.

Public procedures inside the same mechanism cannot call one another.

9.1.3 Private procedures can only be invoked (by means of **call** and **act** statements) inside their own mechanism.

9.1.4 A procedure is active from the time it is invoked to the time it either returns control (explicitly through **exit** or **return** statements or implicitly by reaching the **endproc** statement) or simply terminates.

9.1.5 Synonyms may be declared within a procedure. The scope of such declarations is the procedure itself.

9.2. Example

```
proc PUSH; /* push contents of data stack buffer onto stack*/
    if dstk_ptr = 7 => trap TRAP_RTN.STACK.OVFLOW fi;
    dstk_mem[dstk_ptr] := dstk_buffer;
    dstk_ptr := dstk_ptr + 1
endproc
```

10. FUNCTION DECLARATIONS

<funct> ::= <public funct>|<private funct>
<public funct> ::= **funct** <identifier>;{syn decln;}
   <value clause>{<stmt>;} **value** := <expr>
   endfunct
<private funct> ::= **funct** <identifier> **priv**; {syn decln;}
   <value clause>{<stmt>;} **value** := <expr>
   endfunct
<value clause> ::= **value** : <identifier>|**value** : <type>

10.1. Semantics

10.1.1 An S*A function is similar to a procedure in that it is invoked from public procedures inside other mechanisms (in the case of a public function) or from procedures within the same mechanism (in the case of a private function). However, a function can only be invoked by name as a component of a right-hand-side expression in an assignment statement.

10.1.2 A function returns a value whose type is defined by the specified value clause. The value clause may define the value type explicitly or implicitly, by simply naming a previously defined type identifier. The value returned by a function will always be the last assignment of a value to a variable called **value**.

10.2. Examples

(a) **funct** VAL_AT_DSTK_MEM_AR;
    value : dstk_mem_word;
    value := dstk_mem [dstk_mem_ar]
  endfunct

(b) **funct** VAL_AT_DSTK_MEM_AR;
    value : **tuple** u : **bit**;
                 p : **bit**;
                 a : **bit**;
                 xtype : **seq**[3 .. 0] **bit**;
        endtup;
    value := dstk_mem [dstk_mem_ar]
  endfunct

11. MECHANISMS

<mech class> ::= <mech decln> | <mech type decln>
<mech decln> ::= **mech** <identifier>;<mechanism> **endmech**;
<mechanism> ::= {<mech var decln>;} <proc part>
<mech var decln> ::= <syn decln> | <glovar decln> | <privar decln> |
                   <sync decln> | <const decln> | <chan decln>
<proc part> ::= <proc_funct>{;<proc_funct>}
<proc_funct> ::= <proc> | <funct>
<mech type decln> ::= **type mech** <identifier> =
                    <mechanism> **endtypemech**

11.1. Semantics

11.1.1 A mechanism declaration is the smallest complete unit of architectural description. The hardware entity it describes is termed a mechanism; when there is no ambiguity this term is used to refer to both the construct and the entity it describes.

11.1.2 A mechanism consists of a closed collection of data object and synonym declarations, and one or more procedures and functions that access, operate on, and update these objects. The latter define the procedures' environment.

11.1.3 The only means of accessing a private variable within a mechanism is through one of its procedures (or functions).

11.1.4 A mechanism is active whenever one of its procedures (or functions) is active. It may be activated by initiation statements (see Section 12) or by calls on its public procedures (through **call**, **act**, or **trap**

statements) or functions from public procedures inside other mechanisms.

11.1.5 Mechanisms satisfy the following mutual exclusion rule: a public procedure or function within a mechanism cannot be invoked when the latter is already active.

11.1.6 A mechanism may be shared between systems (see Section 12).

11.1.7 A mechanism represents a particular architectural entity. In contrast a mechanism type declaration defines a class of possible mechanisms, all of which would be exactly identical in structure and function. Given a mechanism type declaration, a collection of identical instances of the type may be defined as a homogeneous system (see Section 12).

11.2. Examples

(a) **mech** INPUT;
    type inst = **seq** [15 .. 0] **bit**;
    glovar input_buffer : **array** [0 .. 7] **of** inst;
    chan bus_A : **seq** [127 .. 0] **bit**;
    sync packet_sent : **bit**(0); /* 0 implies packet not sent*/
    sync ibuffer_empty : **bit**(1); /* 1 implies input buffer empty*/
    sync ibuffer_full : **bit**(0); /* 0 implies input buffer empty*/

    proc IBUFFER_FILL;
      forever do /* wait till input_buffer empty*/
        await ibuffer_empty
          do /* accept int.packet if available*/
            await packet_sent **do** input_buffer := bus_A **od**;
            sig ibuffer_full
          od
      od
    endproc
endmech

(b) **type mech** ALU_DRIVER =
    sync alu_free : **bit**(1);/* 1 implies alu is free*/
    privar opt : opn_pkt; /* holds operation packet*/
    privar result : **seq** [..] **bit**;
    proc DRIVE_ALU;
      forever do
        await alu_free
          do opt := IN.GET_OPN_PKT;
                /* invoke fn.in mech.IN*/
            /* send opt to ALU and receive result from ALU */
            . . . . .

```
      od;
      sig alu_free;
      /* send out "tagged" result to various destinations*/
      . . . . .
   od
  endproc
endtypemech
```

12. SYSTEMS

```
<system> ::= <heterogenous system>|<homogenous system>
<heterogenous system> ::= sys <identifier>;<system comp> endsys
<syst comp> ::= {<sys var decln>;}<mech part>{;<init stmt>}|
                {<heterogenous system>}
<sys var decln> ::= <sync decln>|<glovar decln>|<privar decln>|
                    <syn decln>|<const decln>|<chan decln>
<mech part> ::= <mech decln>{; <mech decln>}
<init stmt> ::= init <proc reference>{,<proc reference>}|
                init <indexed proc reference> . .
                <indexed proc reference>
<indexed proc reference> ::= <proc reference>(<implicit dec integer>)
<homogenous system> ::= sys <identifier>; <mech type decln>;
                        <system set decln> endsys
<system set decln> ::= sys <identifier> : set (<dec int range>)
                       of <identifier>{; <init stmt>}
<dec int range> ::= <implicit dec integer> . . <implicit dec integer>
```

12.1. Semantics

12.1.1 A system encapsulates a set of simpler systems and mechanisms, together with a collection of data objects, and is used to construct architectural descriptions in a hierarchical manner.

12.1.2 A data object declared as private to a system is global with respect to all encapsulated systems and mechanisms.

12.1.3 A data object declared as private to a system can only be accessed by procedures contained in mechanisms defined inside the system.

12.1.4 A heterogenous system describes a set of nonidentical mechanisms.

12.1.5 A homogenous system encapsulates a mechanism *type* T along with a set of mechanisms M = $\{m_1, m_2, \ldots, m_n\}$, each of type T. The members of M are identical in structure and function with the following qualification:

a Let a homogenous system be defined as follows:

> **sys** X;
>   **type mech** Y = . . . **endtypemech**;
>   **sys** z : **set**(1 .. n) **of** Y;
> **endsys**

Then the individual mechanisms in z (each of type Y) are uniquely identified as z(1), z(2), . . . , z(n). Each identifier x defined in the mechanism type declaration is known inside z(i) as x(i)($1 \leq i \leq n$).

12.1.6 Mechanisms within a system may be activated by means of initiation statements. In the case of heterogenous systems, initiation is according to the first form of the **init** statement; in the case of homogenous systems it is according to the second form.

12.2. Examples

(a)   **sys** QM_C;
      sys INSTRUCTION_CYCLE;
        mech INSTRUCTION_FETCH; . . . **endmech**;
        mech INSTRUCTION_DECODE; . . . **endmech**;
        sys INSTRUCTION_EXEC;
          mech FETCH_OPRND; . . . ; **endmech**;
          mech EXECUTE; . . . ; **endmech**;
          mech DEPOSIT_OPRND; . . . ; **endmech endsys**
      endsys
      sys MEMORY;
        mech AUX_MEM; . . . **endmech**
        mech MAIN_MEM; . . . **endmech**
      endsys
      sys INTERRUPT;
        mech INT_HANDLERS; . . . **endmech**
        mech STATE_CHANGE; . . . **endmech**
      endsys
  endsys

(b)   **sys** DRIVER_SYSTEM;
      type mech ALU_DRIVER =
      /* as in example (b), Section 11.2 */
      . . . . .
      endtypemech;
      sys DRIVER : **set** (1 .. n) **of** ALU_DRIVER;
      init DRIVER(1). DRIVE_ALU, . . . , DRIVE(n).DRIVE_ALU;
  endsys

APPENDIX B

SYNTAX AND SEMANTICS OF THE MICROPROGRAMMING LANGUAGE SCHEMA S*

INTRODUCTION

This appendix describes the syntax and semantics of S*. At this time of writing a formal semantics for S* has yet to be constructed, although work is currently under way on developing complete denotational and axiomatic definitions for the schema as reported in Damm (1983) and proof rules and axioms for S*(QM-1) as reported in Wagner (1983) and Wagner and Dasgupta (1983). Thus the semantics presented here is informal. Further, since many of the constructs in S* and S*A are identical, I have avoided repeating the common semantics by simply referring to the appropriate sections of the S*A description (Appendix A) whenever relevant. Such references are of the general form [A,n,..,k] where n,..,k designate the appropriate sections in Appendix A.

1. NOTATION, TERMINOLOGY, AND BASIC VOCABULARY

1.1 Syntactic entities are enclosed within the angular brackets < and >. The notation {x} symbolizes zero or one instance of the entity x. The notations {x}* and {x}+ denote, respectively, zero or more and one or more instances of x.

1.2 The term "S* program" appearing in the following text is an abbreviation for "a (micro)program in any instantiated version of S*." The machine (or microarchitecture) with respect to which an instantiation is done is called the *host* machine (or microarchitecture).

1.3 The basic vocabulary includes letters, digits, special characters, and keywords. The valid keywords are designated explicitly in the syntactic definitions as underscored (or boldface) words; the remainder of the basic vocabulary is as in S*A [A,1.1].

1.4 The syntax of identifiers and integers is as in S*A [A,2,3]. For conciseness, the syntactic category <identifier> will be represented by the symbols I and J. The category <implicit dec integer> (which designates ordinary unsigned decimal integers) will be denoted by the symbols N and M.

1.5 The syntactic categories <test expression> and <right-hand-side expr> are special in that their syntax and semantics are not defined in S*; their precise form and meaning are host machine–dependent and will be defined upon instantiation of S* with respect to a given host machine. These special syntactic categories are denoted in this description by the symbols T and E.

1.6 In an S* program, comments may appear freely, enclosed between the symbols /* and */.

2. PROGRAM AND DATA OBJECT DECLARATIONS

<program> ::= **prog** I; **decl** <data obj decl> **endecl**; <syn decl>;
              {**init** <init list> **endinit**}
              <proc decl>; <main block>
              endprog

<main block> ::= <command>

<data obj decl> ::= <type decl> | <var decl>
              | <const decl> | <pvar decl>
              | <data obj decl>{; <data obj decl>}

<type decl> ::= **type** I = <data type>

<data type> ::= I | **seq**[N..M]**bit** | **bit** | **array**[N..M]**of**<data type>
            {**with** I {,I}*}
           | **tuple** I :<data type>{; I :<data type>}+ **endtup**
           | **stack**[N]**of**<data type>{**with** I {,I}*}
           | **assoc array**[N..M]**of**<data type>

<var decl> ::= **var** I : <data type>{: <data type>}*

<pvar decl> ::= **pvar** I : <data type>{: <data type>}*

SYNONYM, INITIALIZATION, AND PROCEDURE DECLARATION   277

<const decl> ::= **const** I : <const type>
<const type> ::= <int type> (N)M
<int type> ::= **bin**|**dec**|**oct**|**hex**

2.1. Semantics

2.1.1 An S* program is an independent compilable unit.

2.1.2 The (variable) data types in S* are identical to those in S*A. Thus, they satisfy the properties stated in [A,4.1]. Constant data types satisfy the properties given in [A,5.1].

 Context-sensitive restriction: The integer value M given in a constant declaration must (1) be a legal number in the number system specified by <int type> and (2) be representable in N bits.

2.1.3 In an S* program, the declaration of a variable, pseudovariable (**pvar**), or constant data object of some type T denotes the existence in the host machine of a storage device whose abstract properties are given by the properties of the data type T.

2.1.4 *Pragmatic note:* The identifier I of a type, variable, pseudovariable, or constant data object is defined at the time of instantiation. In fact, the entire program component <data obj decl> is defined at instantiation time. The S* programmer cannot *invent* such a declaration; he or she can only *include* one or more such predefined declarations in an S* program.

2.1.5 Pseudovariable declarations denote implicit storage devices in the host machine, the values of which can be altered or tested by the microprogram.

2.1.6 *Pragmatic note:* The two most common uses of a pseudovariable declaration are: (1) to represent a field in a microinstruction word that can be set (implicitly or explicitly) or tested by the microprogram; and (2) to denote hidden data objects in the host machine such that their declarations and use in the executable text enhance the clarity of the S* program.

3. SYNONYM, INITIALIZATION, AND PROCEDURE DECLARATION

<syn decl> ::= **syn**<syn clause>{, <syn clause>}*
    | <syn decl>; <syn decl>
<syn clause> ::= I {: <data type>}+ = <selector>
<selector> ::= I | <selector>[N] | <selector>[N..M]
    | <selector>.I
<init list> ::= <access expr>{, <access expr>}* := <int type>N
    | <access expr>{, <access expr>}* := N
    | <init list>; <init list>

<proc decl> : : = **proc** I {(I {; I}*)};<syn decl>; <command>**endproc**
              | <proc decl>; <proc decl>
<access expr> : : = I | <access expr>.I | <access expr>[I]
                  | <access expr>[I..J] | <access expr>[N]
                  | <access expr>[N..M]
                  | <access expr[I <addop> N]
<addop> : : = + | −

3.1. Semantics

3.1.1 A synonym declaration in S* is identical in syntax and semantics to the static synonym declaration in S*A [A,6.8.10, 6.8.11]

3.1.2 The scope of a synonym declaration is the procedure or program in which it is declared.
Context-sensitive restriction: synonym identifiers must be unique over the entire program.

3.1.3 The assignment statements in an initialization list are executed sequentially, once, at the beginning of program execution—that is, before the start of execution of <main block>.

3.1.4 A procedure declaration describes a subroutine that may be invoked by another procedure within the program, or by the program's <main block>.

3.1.5 Local declarations in a procedure may only be synonym declarations for globally defined data objects. Thus, data object identifiers used in the procedure body are either names of global data objects, globally defined synonyms, or synonyms declared within the procedure. The (optional) parameter list serves merely as a comment.

4. COMMANDS AND STATEMENTS

<command> : : = <stat> | **while**<cond>**do**<command>**od**
              | repeat<command>**until**<cond>
              | **call** I | <jump> | **region**<region list>**endreg**
              | **if**<cond> = > <command>{∥<cond> = >
                <command>}***fi**
              | <label>: <command>
              | **case** I **of**{<case list>: <command>} + **endcase**
              | <command>; <command>
<case list> : : = N | <case list>, <case list>
<jump> : : = **goto** I | **act** I
<stat> : : = skip | <simp stat> | <cocycle> | <stcycle>
           | **do**<stat>{; <stat>}***od** | <stat>; <stat>
           | **do**<stat>{□ <stat>}***od** | <stat> □ <stat>
           | <label>: <stat>

COMMANDS AND STATEMENTS   279

```
<cocycle>    ::= cocycle<simp stat>coend
<stcycle>    ::= stcycle<stat>stend
<simp stat>  ::= <simp ass>|<select>|<simp stat>; <simp stat>
                 |<simp stat>□<simp stat>
                 |do<simp stat>{; <simp stat>}*od
                 |do<simp stat>{□<simp stat>}*od
<select>     ::= if<cond> => <jump>{||| <cond> => <jump>}*fi
<region list> ::= <simp stat>|<cocycle>|<stcycle>
                 |<region list>; <region list>
<cond>       ::= T | <simple test>
<simple test> ::= I[N..M] | I | ˜T | (T) | T<logop>T
<logop>      ::= V | ∧ | ∀ | ∧ | ⊕
<simp ass>   ::= <access expr>{, <access expr>}* := E
                 | push (I[J], <stack access>)
                 | pop (I[J], <stack access>)
<stack access> ::= I | <stack access>.I | <stack access>[I]
                 | <stack access>[N]
                 | <stack access>[I <addop> N]
```

4.1 Semantics

4.1.1 The **while, repeat, case, if,** and **goto** statements have the same semantics as the corresponding constructs in S*A [A,8.3.4, 8.3.5, 8.3.6, 8.2.5].

4.1.2 The procedure **call** statement transfers control to, and initiates execution of, the called procedure I.

4.1.3 The procedure **act** statement initiates execution of the activated procedure I, and terminates the activating procedure.

4.1.4 A **region** statement is simply a specified sequence of **cocycle, stcycle,** and simple statements.

Pragmatic note: The **region** construct is semantically identical to a sequential statement. Its significance is that it allows the microprogrammer to optimize and/or compact code at the source program level. The prescribed flow of control within a **region** must be preserved; hence the compiler must not perform any optimizing transformations that alter the specified ordering of the **region** components.

4.1.5 A **skip** statement has no effect on the machine state. However, the duration of its "execution" is a microcycle.

4.1.6 The sequential statement S1; S2; . . . ; Sn and the parallel statement S1 □ S2 □ . . . □ Sn have identical semantic properties to the corresponding statements in S*A [A,8.3.1, 8.3.2].

4.1.7 The **cocycle** statement begins and terminates in a single microcycle.

4.1.8 The **stcycle** statement begins in a new microcycle. Its termination is not defined by the construct itself. Rather, its termination time is determined by the timing requirements of its constituent statements.

4.1.9 When a <select> statement appears in a **cocycle** (or **stcycle**), the jump, if it takes place, will do so after all the other constituent statements within the **cocycle** (or **stcycle**) have been executed.

4.1.10 The **push** and **pop** statements have the same semantics as the corresponding constructs in S*A [A,4.1.6].

4.1.11 *Context-sensitive restrictions:*
    a There may be at most one occurrence of a <select> statement within a **cocycle** or a **stcycle**.
    b No **cocycle** or **stcycle** may be parallel to another **cocycle** or **stcycle**.
    c There may be no jumps into **proc, while, repeat, stcycle, cocycle, if,** or **case** statements.
    d Jumps out of a procedure are permitted if and only if they are through an **act** statement.
    e The identifier I in the first component I[J] of a **push** or **pop** statement may only be of type **stack**.

REFERENCES

Agrawala, A. K., and Rauscher, T. G. (1976). *Foundations of Microprogramming.* New York: Academic Press.

Alagic, S., and Arbib, M. A. (1978). *The Design of Well-Structured and Correct Programs.* New York: Springer-Verlag.

Alexander, C. (1964). *Notes on the Synthesis of Form.* Cambridge, MA: Harvard University Press.

Alexander, W. G., and Wortman, D. B. (1975). "Static and Dynamic Characteristics of XPL Programs," *Computer,* **8,** 11 (Nov.), 41–46.

Anceau, F., Liddell, P., Mermet, J., and Paycand, C. (1971). "CASSANDRE: A Language to Describe Digital Systems." In J. Tou (Ed.), *Software Engineering,* Vol. 1. New York: Academic Press.

Baer, J-L. (1980). *Computer Systems Architecture.* Potomac, MD: Computer Science Press.

Barbacci, M. R. (1975). "A Comparison of Register Transfer Languages for Describing Computers and Digital Systems," *IEEE Trans. Comput.,* **C-24,** 2.

Barbacci, M. R. (1981). "Instruction Set Processor Specification (ISPS): The Notation and Its Application," *IEEE Trans. Comput.,* **C-30,** 1 (Jan.), 26–40.

Barbacci, M. R., Barnes, G. E., Cattell, R. G., and Siewiorek, D. P. (1978). "The ISPS Computer Description Language." Pittsburgh: Dept. of Computer Science, Carnegie-Mellon University.

Barbacci, M. R., and Northcutt, J. D. (1980). "Application of ISPS, an Architecture Description Language," *J. Digital Systems,* **4,** 3, 221–239.

Barbacci, M. R., and Parker, A. (1978). "Using Emulation to Verify Formal Architecture Descriptions," *Computer* (May), 51–56.

Barnes, G. H., Brown, R. M., Kato, M., Kuck, D. J., Slotnick, D. L., and Stokes, R. A. (1968). "The Illiac IV Computer," *IEEE Trans. Comput.,* **C-17,** 8 (August), 746–757.

Bazjanac, V. (1979). "Architectural Design Theory: Models of the Design Process." In W. R. Spillers (Ed.), *Basic Questions of Design Theory.* Amsterdam: North-Holland Elsevier.

Bell, C. G., and Newell, A. (1971). *Computer Structures: Readings and Examples.* New York: McGraw-Hill.

REFERENCES

Boyd, D. L., and Pizzarello, A. (1978). "An Introduction to the WELLMADE Design Methodology," *IEEE Trans. Soft. Eng.*, **TSE-4,** 4 (July), 276–282.

Brinch Hansen, P. (1977). *The Architecture of Concurrent Programs.* Englewood-Cliffs, NJ: Prentice-Hall.

Brinch Hansen, P. (1981). "The Design of Edison," *Software: Practice and Experience,* **11,** 363–396.

Broadbent, G. (1973). *Design in Architecture.* New York: Wiley.

Brooks, F. P. (1975). *The Mythical Man-Month.* Reading, MA: Addison-Wesley.

Buckle, J. K. (1978). *The ICL 2900 Series.* London: MacMillan.

Bullock, A., and Stallybrass, O., (Eds.) (1977). *The Fontana Dictionary of Modern Thought.* London: Fontana/Collins.

Burston, A. K., Kinniment, D. J., and Kahn, H. (1978). "A Design Language for Asynchronous Logic," *Computer J,* **21,** 4, 347–354.

Campos, I. M., and Estrin, G. (1978). "SARA-Aided Design of Software for Concurrent Systems." In *AFIPS Natl. Comp. Conf. Proc.,* Vol. 47. Arlington, VA: AFIPS Press.

Castan, M., and Organick, E. I. (1982). "M3L: An HLL-RISC Processor for Parallel Execution of FP-Language Programs." In *Proc. 9th Ann. Symp. on Comp. Arch.* New York: IEEE Comp. Soc. Press, 239–247.

Chu, Y. (1965). "An Algol-like Computer Design Language," *Comm. ACM,* **8,** 607–615.

Chu, Y. (1972). *Computer Organization and Microprogramming.* Englewood-Cliffs, NJ: Prentice-Hall.

Clark, D. W., and Strecker, W. D. (1980). "Comments on 'The Case for the Reduced Instruction Set Computer'," *Computer Architecture News,* **8,** 6, October, 34–38.

Courtois, P. J. (1977). *Decomposability.* New York: Academic Press.

Dahl, O-J., and Nygaard, K. (1970). "The Simula 67 Common Base Language." Oslo: Norwegian Computing Center.

Damm, W. (1983). Personal communication.

Dasgupta, S. (1976). "Parallelism in Microprogramming Systems." Ph.D. thesis (Tech. Rept. TR76-4). Edmonton, Alta.: Dept. of Comp. Sc., Univ. of Alberta.

Dasgupta, S. (1977). "Parallelism in Loop-Free Microprograms." In *Information Processing 77* (Proc. IFIP Congress). Amsterdam: North-Holland.

Dasgupta, S. (1978). "Towards a Microprogramming Language Schema." In *Proc. 11th Ann. Workshop on Microprogramming,* 144–153. New York: ACM/IEEE.

Dasgupta, S. (1980a). "Some Implications of Programming Methodology for Microprogramming Language Design." In G. Chroust and J. Mulbacher (Eds.), *Firmware, Microprogramming, and Restructurable Hardware,* 243–252. Amsterdam: North-Holland.

Dasgupta, S. (1980b). "Some Aspects of High-Level Microprogramming," *ACM Comput. Surv.,* **12,** 3 (Sept.), 295–324.

Dasgupta, S. (1981). "S*A: A Language for Describing Computer Architectures." In M. Breuer and R. Hartenstein (Eds.), *Computer Hardware Description Languages and Their Applications,* 65–78. Amsterdam: North-Holland.

Dasgupta, S. (1982). "Computer Design and Description Languages." In M. C. Yovits (Ed.), *Advances in Computers,* Vol. 21. New York: Academic Press.

Dasgupta, S. (1983). "On the Verification of Computer Architecture Using an Architecture Description Language." In *Proc. 10th Annual Symp. on Computer Architecture.* New York: IEEE Comp. Soc. Press.

Dasgupta, S., and Olafsson, M. (1982). "Towards a Family of Languages for the Design and Implementation of Machine Architectures." In *Proc. 9th Annual Symp. on Computer Architecture,* 158–167. New York: IEEE Comp. Soc. Press.

Dasgupta, S., and Tartar, J. (1976). "The Identification of Maximal Parallelism in Straight Line Microprograms," *IEEE Trans. Comput.*, **C-25**, 10 (Oct.), 986–992.

Davidson, S. (1980). "Design and Construction of a Virtual Machine Resource Binding Language." Ph.D. thesis. Lafayette, LA: Dept. of Computer Sc., Univ. of Southwestern Louisiana.

Davidson, S. (1983). "High-Level Microprogramming—Current Usage, Future Prospects." *Proc. 16th Annual Workshop on Microprogramming*, IEEE Comp. Soc. Press, N.Y.

Davidson, S., and Shriver, B. D. (1980). "MARBLE: A High-Level Machine-Independent Language for Microprogramming." In G. Chroust and J. Mulbacher (Eds.), *Firmware, Microprogramming, and Restructurable Hardware*, 253–263. Amsterdam: North-Holland.

Davidson, S., and Shriver, B. D. (1980a). "Firmware Engineering: An Extensive Update." In G. Chroust and J. Mulbacher (Eds.), *Firmware, Microprogramming, and Restructurable Hardware*, 1–40. Amsterdam: North-Holland.

Davidson, S., and Shriver, B. D. (1981). "Specifying Target Resources in a Machine-Independent Higher Level Language." In *Proc. Natl. Comp. Conf.*, 81–85. Arlington, VA: AFIPS Press.

Davis, A. L., and Drongowski, P. J. (1980). "Data Flow Computers: A Tutorial and Survey," Tech. Rept. UUCS-80-109. Salt Lake City: Dept. of Computer Science, Univ. of Utah.

de Bakker, J. (1980). *Mathematical Theory of Program Correctness*. Englewood-Cliffs, NJ: Prentice-Hall International.

Demco, J., and Marsland, T. A. (1976). "An Insight into PDP-11 Emulation." In *Proc. 9th Annual Workshop on Microprogramming*. New York: ACM/IEEE.

Denning, P. J. (1978). "A Question of Semantics," *SIGARCH Comp. Arch. New*, **6**, 8 (April), 16–18.

Dennis, J. B., et al. (1979). "Research Directions in Computer Architecture." In P. Wegner (Ed.), *Research Directions in Software Technology*. Cambridge, MA: MIT Press.

DeWitt, D. J. (1976a). "A Machine-Independent Approach to the Production of Horizontal Microcode." Ph.D. thesis (Tech. Rept. 76 DTA). Ann Arbor: Univ. of Michigan.

DeWitt, D. J. (1976b). "Extensibility—A New Approach for Designing Machine-Independent Microprogramming Languages." In *Proc. 9th Annual Workshop on Microprogramming*. New York: ACM/IEEE.

Dietmeyer, D. L., and Duley, J. R. (1975). "Register-Transfer Languages and Their Translations." In M. A. Breuer (Ed.), *Digital Systems Design Automation*. Potomac, MD: Computer Science Press.

Dijkstra, E. W. (1968). "The Structure of the T.H.E. Multiprogramming System," *Comm. ACM*, **11**, 5, 341–346.

Dijkstra, E. W. (1972). "Notes on Structured Programming." In O-J. Dahl, E. W. Dijkstra, and C. A. R. Hoare, *Structured Programming*. New York: Academic Press.

Dijkstra, E. W. (1976). *A Discipline of Programming*. Englewood-Cliffs, NJ: Prentice-Hall.

Director, S., Parker, A. C., Siewiorek, D. P., and Thomas, D. E. (1981). "A Design Methodology and Computer Aids for Digital VLSI Systems," *IEEE Trans. Circuits and Systems*, **CAS-28**, 7 (July), 634–645.

Ditzel, D. R. (1980). "Program Measurements on a High-Level Language Computer," *Computer*, **13**, 8 (Aug.).

Doeringer, S., Mitten, D. G., and Steinberg, A., (Eds.) (1970). *Art and Technology: A Symposium on Classical Bronzes*. Cambridge, MA: MIT Press.

Doran, R. W. (1979). *Computer Architecture: A Structured Approach*. New York: Academic Press.

Eckhouse, R. H. (1971). "A High-Level Microprogramming Language (MPL)." Ph.D. thesis. Buffalo: Dept. of Computer Science, State Univ. of New York.

Elshoff, J. L. (1976). "An Analysis of Some Commercial PL/1 Programs," *IEEE Trans. Soft. Eng.* (June), 113–120.

Estrin, G. (1978). "A Methodology for Design of Digital Systems—Supported by SARA at the Age of One." In *Proc. AFIPS Natl. Comp. Conf.*, Vol. 47, 313–324. Arlington, VA: AFIPS Press.

Fisher, J. A. (1979). "The Optimization of Horizontal Microcode Within and Beyond Basic Blocks: An Application of Processor Scheduling with Resources." U.S. Dept. of Energy Rept., COO-3077-161. New York: Dept. of Mathematics and Computing, New York University.

Fisher, J. A. (1981). "Trace Scheduling: A Technique for Global Microcode Compaction," *IEEE Trans. Comput.*, **C-30**, 7 (July), 478–490.

Fisher, J. A., Landskov, D., and Shriver, B. D. (1981). "Microcode Compaction: Looking Backward and Looking Forward." In *Proc. AFIPS Natl. Comp. Conf.*, 95–102. Arlington, VA: AFIPS Press.

Fitzpatrick, D. T., et al. (1981). "VLSI Implementation of a Reduced Instruction Set Computer." In H. T. Kung, R. F. Sproull, and G. Steele (Eds.), *VLSI Systems and Computations*. Rockville, MD: Computer Science Press.

Floyd, R. W. (1967). "Assigning Meaning to Programs." In *Mathematical Aspects of Computer Science*, XIX, American Math. Soc.

Flynn, M. J. (1966). "Very High Speed Computing Systems," *Proc. IEEE,* **54** (Dec.), 1901–1909.

Flynn, M. J. (1974). "Trends and Problems in Computer Organizations." In *Information Processing 74* (Proc. IFIP Congress), 3–10. Amsterdam: North-Holland.

Flynn, M. J. (1979). "The Interpretive Interface: Resources and Program Representation in Computer Organization." In D. J. Kuck, D. H. Lawrie, and A. H. Sameh (Eds.), *High-Speed Computer and Algorithm Organization*. New York: Academic Press.

Flynn, M. J. (1980). "Directions and Issues in Architecture and Language," *Computer* (Oct.), 5–22.

Flynn, M. J., and Hoevel, L. W. (1983). "Execution Architecture: The DELTRAN Experiment," *IEEE Trans. Comput.*, **C-32**, 2 (Sept.), 156–174.

Flynn, M. J., and Rosin, R. F. (1971). "Microprogramming: An Introduction and Viewpoint," *IEEE Trans. Comp.*, **C-20**, 7 (July).

Frankel, R. E., and Smoliar, S. W. (1979). "Beyond Register-Transfer: An Algebraic Approach for Architectural Description." In *Proc. 4th Int. Symp. on Computer Hardware Description Languages,* 1–5. Palo Alto, CA: IEEE.

Freeman, P. (1980). "The Context of Design." In P. Freeman and A. Wasserman (Eds.), *Tutorial on Software Design Techniques,* 3rd ed. New York: IEEE Comp. Soc. Press.

Friedman, Y. (1980). *Towards a Scientific Architecture.* Cambridge, MA: MIT Press.

Fuller, S. H., Shaman, P., Lamb, D., and Burr, W. E. (1977a). "Evaluation of Computer Architectures via Test Programs." In *Proc. AFIPS Natl. Comp. Conf.*, Vol. 46, 147–160. Montvale, NJ: AFIPS Press.

Fuller, S. H., Stone, H. S., and Burr, W. E. (1977b). "Initial Selection and Screening of the CFA Candidate Computer Architectures." In *Proc. AFIPS Natl. Comp. Conf.*, Vol. 46, 139–146. Montvale, NJ: AFIPS Press.

Garside, R. G. (1980). *The Architecture of Digital Computers.* Oxford: Clarendon Press.

Glegg, C. L. (1973). *The Science of Design.* Cambridge, England: Cambridge Univ. Press.

Gombrich, E. H. (1972). *The Story of Art,* 13th ed. London: Phaidon Press.

Gries, D. G. (1976a). "An Illustration of Current Ideas on the Derivation of Correctness Proofs and Correct Programs." *IEEE Trans. Soft. Eng.*, **2** (Dec.), 238–244.

Gries, D. G. (1976b). "An Exercise in Proving Parallel Programs Correct." In F. Bauer and K.

Samelson (Eds.), *Language Hierarchies and Interfaces,* Lecture Notes in Comp. Sc., Vol. 46, 57–81. Berlin: Springer-Verlag.

Gries, D. G. (1981). *The Science of Programming.* New York: Springer-Verlag.

Grishman, R. (1980). "The PUMA Project: Computer Design Automation in the University." In *Proc. ACM 1980 Annual Conference,* 490–497.

Grishman, R., and Bogong, S. (1982). "A Preliminary Evaluation of Trace Scheduling for Global Microcode Compaction," Tech. Rept. 42. New York: Dept. of Computer Science, New York Univ.

Guttag, J. (1977). "Abstract Data Types and the Development of Data Structures," *Comm. ACM,* **20,** 6, 396–404.

Hafer, L., and Parker, A. C. (1978). "Register-Transfer Level Digital Design Automation: The Allocation Process." In *Proc. 15th Design Automation Conf.*

Hafer, L., and Parker, A. C. (1982). "Automatic Synthesis of Digital Hardware," *IEEE Trans. Comput.,* C-31, 2 (Feb.), 93–109.

Hagiwara, H., et al. (1980). "A Dynamically Microprogrammable Computer with Low-Level Parallelism," *IEEE Trans. Comput.,* C-29, 7 (July), 577–595.

Hartmann, A. (1977). *A Concurrent Pascal Compiler for Minicomputers,* Lecture Notes in Comp. Sc., Vol. 50. Berlin: Springer-Verlag.

Hayes, J. P. (1978). *Computer Architecture and Organization.* New York: McGraw-Hill.

Hennessey, J., et al. (1982a). "Hardware/Software Tradeoffs for Increased Performance," *Proc. Symp. on Arch. Support for Prog. Lang. & Op. Systems, Comp. Arch. News,* **10,** 2 (March), 2–11.

Hennessey, J., et al. (1982b). "MIPS: A Microprocessor Architecture." In *Proc. 15th Ann. Workshop on Microprogramming,* 17–22. New York: IEEE Comp. Soc. Press.

Hill, F. J., and Peterson, G. R. (1978). *Digital Systems: Hardware Organization and Design.* 2nd ed. New York: John Wiley and Sons.

Hoare, C. A. R. (1969). "An Axiomatic Approach to Computer Programming," *Comm. ACM,* **12,** 10 (Oct.), 576–580, 583.

Hoare, C. A. R. (1972). "Notes on Data Structuring." In O-J. Dahl, E. W. Dijkstra, and C. A. R. Hoare, *Structured Programming.* New York: Academic Press.

Hoare, C. A. R. (1976). "Parallel Programming: An Axiomatic Approach." In F. Bauer and K. Samelson (Eds.), *Language Hierarchies and Interfaces,* Lecture Notes in Comp. Sc., Vol. 46, 11–42. Berlin: Springer-Verlag.

Hoare, C. A. R., and Wirth, N. (1973). "An Axiomatic Definition of the Programming Language Pascal," *Acta Informatica,* **2,** 335–355.

Hobson, R. F., Hannon, P., and Thornburg, G. (1981). "High-Level Microprogramming with APL Syntax." In *Proc. 14th Ann. Workshop on Microprogramming,* 131–142. New York: IEEE Comp. Soc. Press.

Hoevel, L. W. (1974). "Ideal Directly Executed Languages: An Analytical Argument for Emulation," *IEEE Trans. Comput.,* C-23, 8 (Aug.), 759–767.

Hoevel, L. W. (1978). "Directly Executed Languages." Ph.D. thesis. Baltimore, MD: Johns Hopkins Univ.

Hoevel, L. W., and Flynn, M. J. (1977). "The Structure of Directly Executed Languages: A New Theory of Interpretive System Design," Tech. Rept. 130. Stanford, CA: Comp. Syst. Lab., Stanford University.

Hsiao, D. K. (1980). "Data Base Computers." In *Advances in Computers,* M. C. Yovits (Ed.), Vol. 19. New York: Academic Press.

Husson, S. S. (1970). *Microprogramming: Principles and Practices.* Englewood-Cliffs, NJ: Prentice-Hall.

Hwang, K., Su, S. P., and Ni, L. M. (1981). "Vector Computer Architecture and Processing

Techniques." In M. C. Yovits (Ed.), *Advances in Computers,* Vol. 20. New York: Academic Press.

Ibbett, R. N. (1972). "The MU5 Instruction Pipeline," *Computer J.,* **15**.

IBM Corp. (1981). *The IBM System/370 Principles of Operation,* GA 22-77000. White Plains, NY: IBM.

IEEE (1981). *IEEE Trans. on Computers, Special Issue on Microprogramming Tools and Techniques,* **C-30,** 7 (July).

Iliffe, J. K. (1972). *Basic Machine Principles,* 2nd ed. London and New York: MacDonald/Elsevier.

Iliffe, J. K. (1982). *Advanced Computer Design.* London: Prentice-Hall International.

Intel Corp. (1981). *iAPX 432 General Data Processor Architecture Reference Manual.* Santa Clara, CA: Intel.

Jackson, M. A. (1976). *Principles of Program Design.* London: Academic Press.

Janson, H. W. (1969). *History of Art.* Englewood-Cliffs, NJ and New York: Prentice-Hall and Harry N. Abrams.

Johnson, S. C. (1978). "A Portable Compiler: Theory and Practice," *Proc. 5th ACM Symp. on Principles of Prog. Lang.* (Jan.), 97–104.

Johnson, S. C. (1979). "A 32-bit Processor Design," Comp. Sc. Tech. Rept. 80. Murray Hill, NJ: Bell Laboratories.

Jones, C. B. (1980). *Software Development: A Rigorous Approach.* Englewood-Cliffs, NJ: Prentice-Hall International.

Jones, J. C. (1970). *Design Methods: Seeds of Human Futures* (2nd ed., 1980). New York: John Wiley and Sons.

Keedy, J. L. (1978a). "On the Use of Stacks in the Evaluation of Expressions," *SIGARCH Comp. Arch. News,* **6,** 6 (Feb.).

Keedy, J. L. (1978b). "On the Evaluation of Expressions Using Accumulators, Stacks, and Store-Store Instructions," *SIGARCH Comp. Arch. News,* **7,** 4 (Dec.), 24–28.

Keedy, J. L. (1979). "More on the Use of Stacks in the Evaluation of Expressions," *SIGARCH Comp. Arch. News,* **7,** 8 (June), 18–23.

Kernighan, B. W., and Ritchie, D. (1978). *The C Programming Language.* Englewood-Cliffs, NJ: Prentice-Hall.

Kilburn, T., et al. (1969). "A System Design Proposal." In *Information Processing 68* (Proc. IFIP Congress). Amsterdam: North-Holland.

Klassen, A. (1981). "S*(QM-1): An Experimental Evaluation of the High-Level Microprogramming Language Schema S* Using the Nanodata QM-1." M.S. thesis. Edmonton, Alta.: Dept. of Comp. Sc., Univ. of Alberta.

Klassen, A., and Dasgupta, S. (1981). "S*(QM-1): An Instantiation of the High-Level Microprogramming Language Schema S* for the Nanodata QM-1." In *Proc. 14th Ann. Workshop on Microprogramming,* 126–130. New York: IEEE Comp. Soc. Press.

Kleir, R. L., and Ramamorthy, C. V. (1971). "Optimization Strategies for Microprograms," *IEEE Trans. Comput.,* **C-20,** 7 (July), 783–795.

Kogge, P. M. (1981). *The Architecture of Pipelined Computers.* New York: McGraw-Hill.

Kuhn, T. S. (1970). *The Structure of Scientific Revolutions,* 2nd ed. Chicago: Univ. of Chicago Press.

Kung, H. T., and Foster, M. J. (1980). "The Design of Special Purpose VLSI Chips," *Computer,* **13,** 26–40.

Kung, H. T., and Leiserson, C. E. (1980). "Algorithms for VLSI Processors." In C. A. Mead and L. Conway, *Introduction to VLSI Systems,* Section 8.3. Reading, MA: Addison-Wesley.

Lamport, L. (1979). "A New Approach to Proving the Correctness of Multiprocess Programs," *ACM Trans. Prog. Lang. & Syst.*, **1**, 1, 84–97.

Landskov, D., Davidson, S., Shriver, B. D., and Mallett, P. (1980). "Local Microcode Compaction Techniques," *ACM Comp. Surveys*, **12**, 3 (Sept.), 261–294.

Leung, C. K. C. (1979). "ADL: An Architecture Description Language for Packet Communication Systems." In *Proc. 4th Int. Symp. on Computer Hard. Descrip. Lang.*, 6–13. New York: IEEE Comp. Soc.

Leung, C. K. C. (1981). "On a Topdown Design Methodology for Packet Systems." In M. A. Breuer and R. Hartenstein (Eds.), *Computer Hardware Description Languages and Their Applications*, 171–184. Amsterdam: North-Holland.

Levitt, K., and Robinson, L. (1977). "Proof Techniques for Hierarchically Structured Programs." In R. Yeh (Ed.), *Current Trends in Programming Methodology*, Vol. II. Englewood-Cliffs, NJ: Prentice-Hall.

Li, T. (1982). "A VLSI View of Microprogrammed System Design." In *Proc. 15th Ann. Workshop on Microprogramming*, 96–104. New York: IEEE Comp. Soc. Press.

Lim, W.Y-P. (1982). "HIDSL—A Structure Description Language," *Comm. ACM*, **25**, 11 (Nov.), 823–830.

Liskov, B. (1972). "The Design of the Venus Operating System," *Comm. ACM*, **15**, 3 (March), 144–149.

Liskov, B. (1980). "Modular Program Construction Using Abstractions." In D. Bjorner (Ed.), *Abstract Software Specifications*. Berlin: Springer-Verlag.

Ma, P-Y. (1981). "The Design of a Firmware Engineering Tool: The Microcode Compiler." In *Proc. AFIPS Natl. Comp. Conf.*, 87–94. Arlington, VA: AFIPS Press.

Ma, P-Y., and Lewis, T. G. (1980). "Design of a Machine-Independent Optimizing System for Emulator Development," *ACM Trans. on Prog. Lang. & Syst.*, **2.2**, 239–262.

Ma, P-Y., and Lewis, T. G. (1981). "On the Design of a Microcode Compiler for a Machine-Independent High-Level Language," *IEEE Trans. Soft. Eng.* (May), 261–274.

Malik, K., and Lewis, T. G. (1978). "Design Objectives for High-Level Microprogramming Languages." In *Proc. 11th Ann. Workshop on Microprogramming* (Nov.), 154–160. New York: ACM/IEEE.

Mallett, P. (1978). "Methods of Compacting Microprograms." Ph.D. thesis. Lafayette, LA: Dept. of Computer Sc., Univ. of Southwestern Louisiana.

Manna, Z. (1974). *Mathematical Theory of Computation*. New York: McGraw-Hill.

Manville, W. D. (1973). "Microprogramming Support for Programming Languages." Ph.D. thesis, Cambridge: Univ. of Cambridge.

March, L., (Ed.) (1976). *The Architecture of Form*. Cambridge, England: Cambridge Univ. Press.

Marwadel, P. (1981). "A Retargetable Microcode Generation System for a High-Level Microprogramming Language." In *Proc. 14th Ann. Workshop on Microprogramming*, 115–123. New York: IEEE Comp. Soc. Press.

McKeeman, W., Horning, J. J., and Wortman, D. B. (1970). *A Compiler Generator*. Englewood-Cliffs, NJ: Prentice-Hall.

Mead, C. A., and Conway, L. (1980). *Introduction to VLSI Systems*. Reading, MA: Addison-Wesley.

Melliar-Smith, P. M. (1979). "System Specification." In T. Anderson and B. Randell (Eds.), *Computing Systems Reliability*. London: Cambridge Univ. Press.

Microdata Corp. (1970). *Microprogramming Handbook*, 2nd ed. Santa Ana, CA: Microdata.

Morris, D., and Ibbett, R. N. (1979). *The MU5 Computer System*. London: MacMillan.

Myers, G. J. (1977). "The Case Against Stack-Oriented Instruction Sets," *SIGARCH Comp. Arch. News*, **6**, 3 (Aug.).

Myers, G. J. (1978a). Letter to the editor, *SIGARCH Comp. Arch. News*, **6**, 8 (April).

Myers, G. J. (1978b). "The Evaluation of Expressions in a Storage-Storage Architecture," *SIGARCH Comp. Arch. News*, **6**, 9 (June).

Myers, G. J. (1982). *Advances in Computer Architecture.* New York: John Wiley and Sons (Wiley-Interscience).

Nanodata Corp. (1979). *The QM-1 Hardware Level User's Manual,* revised edition. Williamsburg, NY: Nanodata.

Neuhauser, C. J. (1980). "Analysis of the PDP-11 Instruction Stream," Tech. Rept. 183. Stanford, CA: Comp. Systems Lab., Stanford University.

Olafsson, M. (1981). "The QM-C: A C-Oriented Instruction Set Architecture for the Nanodata QM-1." M.S. thesis (Tech. Rept. TR81-11). Edmonton, Alta: Dept. of Computing Science, Univ. of Alberta.

Organick, E. I. (1973). *Computer System Organization: The B5700/B6700 Series.* New York: Academic Press.

Owicki, S., and Gries, D. G. (1976). "An Axiomatic Proof Technique for Parallel Programs," *Acta Informatica*, **6**, 319–340.

Parker, A. C., and Wallace, J. J. (1979), "SLIDE—An I/O Hardware Descriptive Language." In *Proc. 4th Int. Symp. on Comp. Hardware Description Languages*, 82–88. New York: IEEE Comp. Soc.

Parker, A. C., and Wallace, J. J. (1981). "An I/O Hardware Description Language," *IEEE Trans. Comput.*, **C-30**, 6 (June), 423–428.

Parker, A. C., et al. (1979). "The CMU Design Automation System: An Example of Automated Data Path Design." In *Proc. 16th Ann. Design Automation Conf.* (June), 73–80.

Parnas, D. L. (1972). "On the Criteria to be Used in Decomposing Systems into Modules," *Comm. ACM*, **5**, 12 (Dec.), 1053–1058.

Parnas, D. L. (1974), "On a Buzzword: Hierarchical Systems." In *Information Processing 74* (Proc. IFIP Congress). Amsterdam: North-Holland.

Parnas, D. L., and Darringer, J. A. (1967). "SODAS and a Methodology for System Design." In *AFIPS Fall Jt. Comp. Conf.*, 449–474. Vol. 31, Montvale, NJ: AFIPS Press.

Pattee, H. H., (Ed.) (1973). *Hierarchy Theory: The Challenge of Complex Systems.* New York: George Braziller.

Patterson, D. A. (1976). "STRUM: Structured Programming System for Correct Firmware," *IEEE Trans. Comput.*, **C-25**, 10 (Oct.), 974–985.

Patterson, D. A. (1977). "Verification of Microprograms." Ph.D. thesis (Tech. Rept. UCLA-Eng-7707). Los Angeles, CA: Dept. of Computer Sc., Univ. of California.

Patterson, D. A. (1978). "An Approach to Firmware Engineering." In *Proc. AFIPS Natl. Comp. Conf.*, 643–674. Arlington, VA: AFIPS Press.

Patterson, D. A. (1981). "An Experiment in High-Level Language Microprogramming and Verification," *Comm. ACM*, **24**, 10 (Oct.), 699–709.

Patterson, D. A., and Ditzel, D. (1980). "The Case for the Reduced Instruction Set Computer," *SIGARCH Comp. Arch. News*, **8**, 6 (Oct.), 25–33.

Patterson, D. A., and Piepho, R. S. (1982). "RISC Assessment: A High-Level Language Experiment." In *Proc. 9th Annual Symp. on Comp. Arch.*, 3–8. New York: IEEE Comp. Soc. Press.

Patterson, D. A., and Sequin, C. (1981). "RISC I: A Reduced Instruction Set VLSI Computer." In *Proc. 8th Annual Symp. on Comp. Arch.* New York: IEEE Comp. Soc. Press.

Pevsner, N. (1963). *An Outline of European Architecture.* New York: Penguin.

Piloty, R., et al. (1980a). "CONLAN—A Formal Construction Method for Hardware Description Languages: Basic Principles." In *Proc. AFIPS Natl. Comp. Conf.*, Vol. 49, 209–217. Arlington, VA: AFIPS Press.

Piloty, R., et al. (1980b). "CONLAN—A Formal Construction Method for Hardware Description Languages: Language Derivation." In *Proc. AFIPS Natl. Comp. Conf.*, Vol. 49, 219–227. Arlington, VA: AFIPS Press.

Piloty, R., et al. (1980c), "CONLAN—A Formal Construction Method for Hardware Description Languages: Language Application." In *Proc. AFIPS Natl. Comp. Conf.*, Vol. 49, 229–236. Arlington, VA: AFIPS Press.

Piloty, R., et al. (1981). "CONLAN Draft Report," unpublished working document. Pittsburgh: Computer Science Dept., Carnegie-Mellon University.

Popper, K. R. (1965). *Conjectures and Refutations: The Growth of Scientific Knowledge.* New York: Harper and Row.

Popper, K. R. (1968). *The Logic of Scientific Discovery.* New York: Harper and Row.

Popper, K. R. (1972). *Objective Knowledge: An Evolutionary Approach.* Oxford: The Clarendon Press.

Radin, G. (1982). "The 801 Minicomputer," *Proc. Symp. on Arch. Support for Prog. Lang. & Op. Syst., SIGARCH Comp. Arch. News,* **10,** 2 (March), 39–47.

Ramamoorthy, C. V., and Li, H. F. (1977). "Pipeline Architecture," *ACM Comp. Surv.,* **9,** 1 (March), 61–102.

Ramamoorthy, C. V., and Tsuchiya, M. (1974). "A High-Level Language for Horizontal Microprogramming," *IEEE Trans. Comput.,* **C-23,** 8 (Aug.), 791–801.

Randell, B., and Russell, L. J. (1964). *Algol 60 Implementation.* New York: Academic Press.

Reddi, S. S., and Feustal, E. A. (1976). "A Conceptual Framework for Computer Architectures," *ACM Comp. Surv.,* **8,** 2 (June), 277–300.

Rideout, D. J. (1981a). "An Application of a Microcode Compaction Technique to the Nanodata QM-1 Nanoarchitecture." M.S. thesis. Edmonton, Alta: Dept. of Computing Science, Univ. of Alberta.

Rideout, D. J. (1981b). "Considerations for Local Compactions of Nanocode for the Nanodata QM-1." In *Proc. 14th Ann. Workshop on Microprogramming,* 205–214. New York: IEEE Comp. Soc. Press.

Rosen, S. (1969). "Electronic Computers: A Historical Survey," *ACM Comp. Surv.,* **1,** 1 (March), 7–36.

Rosin, R. F., Frieder, G., and Eckhouse, R. H. (1972), "An Environment for Research in Microprogramming and Emulation," *Comm. ACM,* **15,** 8 (Aug.).

Russell, R. D. (1978). "The PDP-11: A Case Study of How Not to Design Condition Codes." In *Proc. 5th Ann. Symp. on Comp. Arch.,* 190–194. New York: ACM/IEEE.

Salisbury, A. B. (1976). *Microprogrammable Computer Architectures.* New York: Elsevier.

Shibayama, K., et al. (1980). "Performance Evaluation and Improvement of a Dynamically Microprogrammable Computer with Low-Level Parallelism." In *Information Processing 80* (Proc. IFIP Congress), 181–186. Amsterdam: North-Holland.

Siewiorek, D. P., Bell, C. G., and Newell, A. (1982). *Computer Structures: Principles and Examples.* New York: McGraw-Hill.

Simon, H. A. (1962). "The Architecture of Complexity," *Proc. Amer. Phil. Soc.,* **106** (Dec.), 467–482.

Simon, H. A. (1975). "Style in Design." In C. Eastman (Ed.), *Spatial Synthesis in Computer Aided Building Design.* New York: John Wiley and Sons.

Simon, H. A. (1981). *Sciences of the Artificial,* 2nd ed. Cambridge, MA: MIT Press.

Simon, H. A., and Ando, A. (1961). "Aggregation of Variables in Dynamic Systems," *Econometrica*, **29**, 111–138.

Sint, M. (1980). "A Survey of High-Level Microprogramming Languages." In *Proc. 13th Annual Workshop on Microprogramming*, 141–153. New York: IEEE Comp. Soc. Press.

Sitton, W. G. (1973). "Strategies for Microprogram Optimization." Ph.D. thesis. Edmonton, Alta.: Dept. of Computing Science, Univ. of Alberta.

Smith, C. S. (1970). Discussion 1.1. In S. Doeringer, D. G. Mitten, and A. Steinberg (Eds.), *Art and Technology: A Symposium on Classical Bronzes*. Cambridge, MA: MIT Press.

Spillers, W. R., (Ed.) (1979). *Basic Questions of Design Theory*. Amsterdam: North-Holland Elsevier.

Steadman, P. (1979). *The Evolution of Designs*. Cambridge, England: Cambridge Univ. Press.

Stritter, S., and Tredennick, N. (1978). "Microprogrammed Implementation of a Single Chip Microprocessor." In *Proc. 11th Ann. Workshop on Microprogramming*, 8–16. New York: ACM/IEEE.

Sumner, F. (1974). "MU5—An Assessment of the Design." In *Information Processing 74* (Proc. IFIP Congress). Amsterdam: North-Holland.

Suppe, F., (Ed.) (1977). *The Structure of Scientific Theories*, 2nd ed. Urbana, IL: Univ. of Illinois Press.

Sykes, J. B., (Ed.) (1976). *The Concise Oxford Dictionary*, Sixth Edition. Oxford: Clarendon Press.

Tan, C. J. (1978). "Code Optimization Techniques for Microcode Compilers." In *Proc. AFIPS Natl. Comp. Conf.*, 649–664. Arlington, VA: AFIPS Press.

Tanenbaum, A. (1978). "Implications of Structured Programming for Machine Architectures." *Comm. ACM*, **21**, 3 (March), 237–246.

Thomas, D. E., and Siewiorek, D. P. (1981). "Measuring Designer Performance to Verify Design Automation Systems," *IEEE Trans. Comput.*, **C-30**, 1 (Jan.), 48–61.

Tomita, S., et al. (1977). "Hardware Organization of a Low-Level Parallel Processor." In *Information Processing 77* (Proc. IFIP Congress), 855–860. Amsterdam: North-Holland.

Torkoro, M., Tamura, E., and Takizuka, T. (1981). "Optimization of Microprograms," *IEEE Trans. Comput.*, **C-30**, 7 (July), 491–504.

Treleavan, P. C., (Ed.) (1980). *Proc. Joint SRC/Univ. of Newcastle-upon-Tyne Workshop on VLSI: Machine Architecture and Very High Level Languages*. Newcastle-upon-Tyne, England: Computing Laboratory, Univ. of Newcastle-upon-Tyne.

Treleavan, P. C., Brownbridge, D. R., and Hopkins, R. P. (1982). "Data Driven and Demand Driven Computer Architecture," *ACM Comp. Surv.*, **14**, 1 (March), 93–144.

Varian Data Machines (1975). *Varian Microprogramming Guide*. Irvine, CA: Varian.

Wagner, A. (1983). "Verification of S*(QM-1) Microprograms." M.S. thesis. Edmonton, Alta.: Dept. of Computing Science, Univ. of Alberta.

Wagner, A., and Dasgupta, S. (1983). "Proof Rules for a Microprogramming Language Based on S* and Its Application to Verification." Submitted to *Proc. 16th Ann. Workshop on Microprogramming*.

Whyte, L. L., Wilson, A. G., and Wilson, D., (Eds.) (1969). *Hierarchical Structures*. New York: American Elsevier.

Wilkes, M. V. (1973). "The Use of a Writable Control Store for Multiprogramming." In *Proc. 6th Ann. Workshop on Microprogramming*. New York: ACM.

Wilkes, M. V. (1975). *Time-Sharing Computer Systems*, 3rd ed. MacDonald and Jane/American Elsevier.

Wilkes, M. V., and Needham, R. M. (1980). *The Cambridge CAP Computer and Its Operating System*. New York: North-Holland.

REFERENCES

Wilner, W. T. (1972). "Design of the Burroughs B1700." In *Proc. Fall Jt. Comp. Conf.*, 489–497. Montvale, NJ: AFIPS Press.

Wirth, N. (1971). "Program Development by Stepwise Refinement," *Comm. ACM*, **14**, 4 (April), 221–227.

Wirth, N. (1973). *Systematic Programming: An Introduction.* Englewood-Cliffs, NJ: Prentice-Hall.

Wirth, N. (1981). "Lilith: A Personal Computer for the Software Engineer." In *Proc. 5th Int. Conf. on Soft. Eng.*, 2–15. New York: IEEE Comp. Soc. Press.

Wood, G. (1978). "On the Packing of Micro-Operations into Microinstruction Words." In *Proc. 11th Ann. Workshop on Microprogramming*, 51–55. New York: ACM/IEEE.

Wood, G. (1979). "The Computer-Aided Design of Microprograms." Ph.D. thesis (Tech. Rept. CST-5-79). Edinburgh: Dept. of Computer Science, Edinburgh Univ.

Wulf, W. A. (1981). "Compilers and Computer Architecture," *Computer*, **14**, 7 (July), 41–47.

Yau, S. S., Schowe, A., and Tsuchiya, M. (1974). "On Storage Optimization for Horizontal Microprograms." In *Proc. 7th Ann. Workshop on Microprogramming*, 98–106. New York: ACM.

Zemanek, H. (1980). "Abstract Architecture." In D. Bjorner (Ed.), *Abstract Software Specification.* New York: Springer-Verlag.

Ziegler, S., et al. (1981). "Ada for the Intel 432 Microcomputer," *Computer*, **14**, 6 (June), 47–56.

Zimmermann, G. (1980). "MDS—The Mimola Design Method," *J. Digital Sys.*, **4**, 3, 337–369.

Zurcher, W., and Randell, B. (1968). "Iterative Multilevel Modeling: A Methodology for Computer System Design." In *Information Processing 68* (Proc. IFIP Congress), D138–D142. Amsterdam: North-Holland.

INDEX

Abstract data types, 20, 28, 38, 39
Abstraction:
    data, 38, 48, 49, 50, 51, 151–152
    levels of, 2, 21–22, 35, 40, 123–124, 132–134
    procedural, 152
    semaphore-like, 56
Abstract machines, 20, 119, 195, 216
ADA, 120
ADL, 25–26, 33, 37, 38, 39, 61, 188–189
Agrawala, A.K., 10, 118, 122, 125
AHPL, 26, 38, 39
Alagic, S., 84, 139, 215
Alexander, C., viii, 17, 63
Alexander, W., 200, 201, 203, 210
Algol 60, 38, 221
Algorithms, 63
    asymptotic efficiency of, 63
Anceau, F., 32, 39
Ando, A., 26, 38
APL, 26, 38
Applied arts, 17
Arbib, M.A., 84, 139, 215
Architectural integrity, 8–10
Architectural languages, x, 24
Architectural modules, 49
Architectural style, 23
Architectural systems, 42, 49
Architecture:
    of buildings, 11, 17
    of computers, 1–10
    as craft, 11–15, 214
    as design discipline, 40
    levels of, 4–8, 46
Arithmetic logic units, 59
Artificial intelligence, 31

Assertions, 70, 77, 238–245
    notation for, 235
Assignment statement, 46, 93, 96–98, 168–169, 202–207
Associative store, 49, 89, 90
Asynchronous systems, 55–58, 85–86
Auxiliary variables, 94–95
Axiom of assignment, 133
Axiomatic proof rules, 70–73, 75, 93–94

Baer, J.L., 6, 10, 190, 248
Barbacci, M.R., x, 10, 38, 39, 61
Barnes, G.H., 221
Bell, C.G., 1, 10, 36
Binding, of resources, 121–122, 125
Bit-slice logic, 48
Bogong, S., 114
Boyd, D.L., 20, 23
Brinch Hansen, P., 120, 126, 224, 250
Broadbent, G., 17
Broadcasting, of values in assignment statements, 133
Brooks, F.P., 8, 10
Brownbridge, D. R., 39
Bullock, A., ix, 19, 23
Burr, W.E., 10
Burroughs B1700, 208, 209
Burroughs B6700/7700 series, 179, 190, 204
Burroughs D Machine, 114, 119
Burroughs Scientific Processor, 178
Burston, A.K., 25, 39

Cal Data Processor, 118
Campos, I.M., 23
CAP, 3, 7, 15–16, 17, 221
Capabilities, 14–16, 49

293

INDEX

Capability principle, 3, 14–16
Capability store, 49, 50
CASSANDRE, 32, 39
Castan, M., 180, 191
CDC STAR-100, 178
CDL, 25, 32, 38, 39
Channels, in S*A, 45
Chu, Y., 25, 38, 39
Clark, D.W., 191
CMU Design Automation System, 182–183, 191
Cobegin statement, 93
Cocycle statement, 134–136
Complexity, of systems, viii, 1–4, 10, 142
Computer:
   complex instruction set, 180
   generations of, 175–176
   reduced instruction set, 180
Computer architecture:
   accumulator-stack based, 204–207
   capability based, 14–16
   conceptual integrity, 8–10, 11
   crisis in, viii
   definition, 1–10
   as a design discipline, vii–ix, 185, 192, 212–214, 246
   esthetics of, 81
   high level language, 193
   language directed, 187, 193, 200, 216
   levels of, 4–8, 46
   microprogrammed, 63
   pipelined, 13–14
   problem-directed, 187–188
   of the QM-1, 142–158
   register oriented, 204–207
   as science, 213
   software directed, 193
   stack based, 204–207, 221, 222
   storage-to-storage, 204–207
   a symbolic view, ix, 247
   systolic style, 184
   von Neumann model, 31
Computer endoarchitecture:
   definition, 6
   examples, 7, 13, 62–66, 224–225, 227–229
Computer exoarchitecture:
   definition, 6
   examples, 7, 49, 57, 218, 221–224, 225–226
Computer microarchitecture:
   definition, 7
   examples, 7, 117–119, 142–158
Computers:
   Burroughs B1700, 208, 209
   Burroughs D Machine, 114, 119
   Burroughs Scientific Processor, 178

Burroughs 6700/7700, 179, 190, 204
Cal Data Processor, 118
CAP, 3, 7, 15–16, 17, 221
CDC STAR-100, 178
CRAY-I, 178
EM-1, 200
Hewlett Packard HP2115, 114
IBM 801, 180
IBM 7090, 197, 198, 199, 204
IBM System/360, 125, 197, 198, 199, 204
IBM System/370, 8, 184, 196, 204
ILLIAC IV, 178, 221
Intel iAPX-432, 120, 179, 190
Intel 8086, 204
Microdata 1600, 118
Microdata 3200, 118, 122
MIPS, 180
Motorola 68000, 221
M3L, 180
MU5, 13–14, 17, 49, 85–86, 178
PDP-10, 198
PDP-11, 9, 49, 114, 140, 170, 184, 185, 198, 204
PUMA, 114
QA-1, 111, 125
QM-C, 187, 200
QM-1, 9, 41, 49, 61, 114, 118–119, 125, 185, 216
RISC-I, 180, 187, 202
SWARD, 57, 58, 61, 187, 224
Texas Instruments ASC, 178
Varian 75, 117, 125, 139
VAX, 202
Concurrency:
   in S*, 136–137
   in S*A, 51, 55–57
Concurrent Pascal, 38, 120, 126
Concurrent Pascal machine, 224
CONLAN, 33, 38, 39, 61
Control store, 64
Conway, L., 184, 190, 191
Courtois, P.J., 194, 195
C programming language, 187, 189, 200–202, 216
CRAY-I, 178
Critical regions, 51, 56

Dahl, O-J, 120
Damm, W., 260
Darringer, J.A., 22
Dasgupta, S., 10, 32, 38, 39, 61, 91, 114, 125, 127, 139, 141, 159, 172, 214, 249, 260
Data abstraction, 38, 48, 49, 50, 51, 210
Database computers, 221

INDEX

Data flow computers, 31, 39, 221
Data objects:
  scope rule in S*, 128
  scope rule in S*(QM-1), 159
Data types, 43–45, 67–68, 194–196
Davidson, S., 38, 39, 120, 121, 125, 126
Davis, A.L., 221
DDL, 25, 32, 39
deBakker, J., 84
Decomposability, 143, 194, 219–220
Demco, J., 140, 184
Denning, P.J., 211
Dennis, J.B., 5, 10, 39, 211
Descriptions:
  behavioral, 32–37
  functional, 28–30, 39
  levels of, 1–2, 5, 33
  modular, 48–50, 69
  nonprocedural, 30–32
  operational, 28–30, 39, 60
  procedural, 30–32
  structural, 32–37, 60
Design:
  analysis in, 215
  automation, 33, 247
  bottom up, 19, 22
  correctness of, vii, 16, 18, 64, 73–80, 92, 98, 234–245
  craft stage of, 12, 214
  creativity in, 221
  decisions, 19, 193–197, 221, 224
  by drawing, 12
  esthetics of, 81
  evolution of, 16–17
  formal, 14, 18–23, 62, 66–70, 86, 214, 226–234, 246
  as heuristic search, 180–182, 217, 246
  informal, 11–16, 18, 22–23, 64–66, 85, 219–226, 246
  inside out, 19
  as an iterative process, 227
  knowledge base, 214, 246
  mathematization of, viii
  methodology, ix, 19–22, 50, 120, 248
  methods, 19
  outside in, 19, 22, 183, 218, 224
  by stepwise refinement, 183, 215–216, 218, 226
  style, 180–184, 214
  as synthesis, 215
  theory, viii
  top down, 19, 22
  validation, 12, 16, 18
DeWitt, D.J., 38, 39, 120, 122, 126

Dietmeyer, D.L., 24, 38, 39
Digital systems, 31
Dijkstra, E.W., viii, 77, 81, 84, 195, 215
Directly executable languages, 200, 223
Director, S.W., 190, 191
Distributed systems, viii
Ditzel, D., 180, 191, 208, 223
Doeringer, H., 190
Doran, R.W., 190
Drongowski, P., 221
Duley, J., 24, 38, 39

Eckhouse, R.H., 10, 38
Edison, 250
Elshoff, J.L., 200, 203
EM-1, 200
EMPL, 38, 120–121, 126
Emulation:
  direct, 140, 170
  hybrid, 140
  as implementation style, 184–185, 217–218, 226
  indirect, 140
Environment:
  inner, 192
  outer, 192, 197, 219
  virtual, 192
Estrin, G., 22, 23, 36, 215

Family of design languages, ix, 40–43, 110, 120, 123
Feustal, E.A., 5, 10
Fisher, J.A., 115, 126
Fitzpatrick, D.T., 191
Flow structure, 86
Floyd, R.W., 70, 84
Floyd-Hoare proof method, 71–73
Flynn, M.J., 154, 177, 190, 191, 197, 198, 199, 200, 204
FORTRAN, 38
Frankel, R.E., 30, 38, 39
Freeman, P., 19, 23, 218
Frieder, G., 10
Friedman, Y., 17
F type instructions, 197–199
Fuller, S.H., 5, 10

Garside, R.G., 248
Gibson mix, 198
Glegg, C.L., 17
Global variables in S*A, 44, 53–55
Gombrich, E.H., 190
Goto statement, proof rule for, 139
Graph model of computation, 22

Gries, D.G., 77, 84, 92, 93, 109
Grishman, R., 114
Guttag, J., 30, 38

Hager, L., 191, 247
Hagiwara, H., 111, 125
Hardware description languages, 24, 30
Hartmann, A., 126
Hayes, J.P., 248
Hennessey, J., 180, 191
Hewlett-Packard HP2115, 114
Hierarchical design, 195–196, 217–218, 220
Hierarchical modeling, 21
Hierarchy theory, 1–4, 10, 143, 194
High level microprogramming, 110
  problems encountered in, 111–119
Hill, F.J., 26, 38, 39
Hoare, C.A.R., viii, 70, 84, 109, 250
Hobson, R.F., 139
Hoevel, L., 200, 223
Hopkins, R.P., 39
Huffman encoding, 186, 208–209
Husson, S.S., 38, 125
Hwang, K., 190

IBM 801, 180
IBM 7090, 197, 198, 199, 204
IBM System/360, 125, 197, 198, 199, 204
IBM System/370, 8, 184, 194, 204
Iliffe, J.K., 193, 211
ILLIAC IV, 178, 221
Implementation, levels of, 33
Imprecise interrupt, 9, 10
Indivisibility of actions, 93, 96–97, 136
Information hiding, 3, 6, 48, 50, 52, 215, 229–230
Instantiated language, 41, 42, 122–125
Instantiation of S*, 141, 158–171
Instruction buffer unit, 85
Instruction cycle, 177
Instruction format, interaction with storage organization, 196
Instruction pipeline, 13, 86, 178
Instruction types:
  frequency analysis of, 186–187, 199–210
  F Type, 197–199
  M Type, 197–199
  P Type, 197–199
  relation with data types, 195–196
  relation with instruction formats, 195
Intel 8080, 204
Intel iAPX-432, 120, 179, 190
Intellectual management of complexity, 11

Interference, freedom from, 94, 100–104, 136
ISPS, x, 37, 38, 61, 119, 248

Jackson, M.A., 185
Janson, H.W., 190
Johnson, S.C., 200, 201
Jones, C.B., 84
Jones, J.C., 12, 16

Kahn, H., 25, 39
Keedy, J.L., 200, 204–207
Kernighan, B.W., 201, 216
Kilburn, T., 91
Kinniment, D.J., 25, 39
Klassen, A., 39, 61, 140, 141, 159, 172
Kleir, R.L., 125
Kogge, P.M., 10, 190
Kuhn, T.S., 213, 247
Kung, H.T., 184
K vector, 119, 143

Lamport, L., 109
Landskov, D., 114, 125, 141, 172
Language:
  extensible, 120–122
  instantiated, 41, 42, 122–125
  schema, 41, 42, 110, 122–125, 127
Language examples:
  ADA, 120
  ADL, 25–26, 33, 37, 38, 39, 61, 188–189
  AHPL, 26, 38, 39
  Algol 60, 38, 221
  APL, 26, 38
  C, 187, 189, 200, 201
  CASSANDRE, 32, 39
  CDL, 25, 26, 32, 38, 39
  Concurrent Pascal, 38, 120
  CONLAN, 33, 38, 39
  DDL, 25, 32, 37
  Edison, 250
  EMPL, 38, 120–121, 126
  FORTRAN, 38
  ISPS, 37, 38, 61, 119
  MARBLE, 38, 120–122, 126
  MPL, 38
  Pascal, 70–71, 93, 114, 121, 187, 200, 202
  PL/1, 38, 201
  S*, 38, 40–41, 61, 110, 125, 127–140
  S*A, 32, 38–39, 41, 49, 50, 61, 64, 110, 119, 226, 231
  SAL, 200–201
  SIMULA, 120
  SLIDE, 32, 38, 45, 61

INDEX

SPL, 208
S*(QM-1), 61, 140, 158–171, 231
STRUM, 38, 119
TRANSLANG, 114
VMPL, 119, 126
XPL, 201
Languages:
  Algol like, 38
  Architectural, x, 24
  Behavioral, 25, 32–37
  Family of, ix, 40–43, 110, 120, 123
  Functional, 25, 28–30
  Hardware description, 24, 30
  Kin, 40, 110, 231
  Microprogramming, x, 25, 38, 39, 42, 110, 119–122, 171
  Nonprocedural, 25, 30–32
  Operational, 25, 28–30
  Procedural, 25, 30–32, 41
  Programming, 24, 28, 30, 37, 47
  Register transfer, 25, 33, 35, 38, 46
  Structural, 25, 32–37
Large scale integration (LSI), viii, 48
Leiserson, C.E., 184
Leung, C.K.C., 25, 38, 39, 59, 61, 188
Levels of abstraction, 2, 21–22, 25
Levels of architecture, 4–8
Levels of construction, 2
Levels of description, 1–2, 5
Levitt, K., 39
Lewis, T.G., 114, 119, 125, 126, 172
Li, H.F., 10, 190
Li, T., 117, 120
Liskov, B., 152, 195
Local microcode compaction, 114, 115, 125, 141
Loop invariant, 71, 78, 99–100, 240–244
Lost wax casting, 176

Ma, P-Y.W., 114, 126, 172
Malik, K., 119, 125, 126
Mallett, P.W., 126
Manna, Z., 84
Manville, W.D., 84
MARBLE, 38, 120–122, 126
March, L., 17
Marsland, T.A., 140, 185
Marwadel, P., 191
Mead, C.A., 184, 190, 191
Mechanism, for synchronization, 38
Mechanism construct in S*A, 38, 42, 44, 50–55
Mechanism types in S*A, 58
Melliar-Smith, P.M., 28, 39

Method, 18
Methodology of design, ix, 19–22, 50, 120, 248
Microcode loader, 63–64
Microcycle, 134–136, 156
Microdata 1600, 118
Microdata 3200, 118, 122
Microinstruction, 64, 111, 117, 156
Microoperation, 111, 117, 120–122
Microprogram:
  compaction, 112–115, 172
  global compaction, 114–115
  local compaction, 114–115, 125, 141
  optimization, 113–114, 121, 125, 137
  portability, 115–117
  store, 64, 118, 124
  verification, 119, 123
Microprogramming:
  high level, 110, 119–120, 123–125, 171
  languages, x, 25, 38, 39, 42, 110, 119–122, 171
  machine dependent, 121–125
  machine independent, 115, 121–125
Microprograms:
  compilers for, 112
  for emulation, 140
  horizontal, 113, 217
  machine-specificity of, 115–119, 120–122
  optimality of, 112–115
  parallelism in, 111–112, 134–137
  vertical, 113, 125, 217
Migration of functions, viii
MIMD machines, 177
MIMOLA design method, 188, 191
MIPS, 180
MISD machines, 177
Modularization, 48, 69, 137–138, 143, 188, 194
Monitors, 38
Morris, D., 13, 17, 85, 91
Motorola 68000, 221
MPL, 38
M ratio, 198–199, 204
M3L, 180
M type instructions, 197–199
MU5, 13–14, 17, 49, 85–86, 178
Myers, G.J., 5, 8, 10, 52, 57, 61, 186, 187, 191, 192, 193, 200, 204–208, 211

Nanoarchitecture, 9, 141, 217
Nanocontrol, 143, 156–158
Nanodata QM-1, 9, 41, 49, 61, 114, 118–119, 125, 140–141, 142–158
Nanoinstruction, 156–158
Nanoprogram, 217

Nanostore, 118, 140–141
Needham, R.M., 3, 10, 15, 17, 221
Network:
 of computers, viii, 48
 of interconnections, 33
 of microprogrammable processors, 111
Neuhauser, J., 200
Newell, A., 1, 10, 36
NF ratio, 198–199
Northcutt, J.D., 61
Nygaard, K., 120

Occam's razor, 81
Olafsson, M., 39, 61, 185, 187, 189, 191, 200, 201, 203, 209, 214, 223, 249
Operation code, 65, 73–74
Optimization of microprograms, 113–114, 121, 125, 137
Organick, E.I., 180, 190, 191
Owicki, S., 92, 93, 109
Owicki-Gries proof method, 93–95

Paradigm shifts, 247–248
Paradigms in science, 213
Parallelism, 42, 55–57, 95–96
Parallel statements, 42, 75, 79–80, 93, 134–137
Parker, A.C., 32, 38, 39, 61, 191, 247
Parker, R.A., 10
Parnas, D.L., ix, 3, 10, 22, 33, 39, 215, 227
Parnas modules, 39
Pascal, 70, 71, 93, 114, 121, 187, 200, 202
Pattee, H.H., 10
Patterson, D.A., 38, 39, 114, 119, 126, 180, 187, 191, 200, 202, 203, 209, 223
PDP-10, 198
PDP-11, 9, 49, 114, 140, 170, 184–185, 198, 204
Performance characteristics, 87
Peterson, G., 26, 38, 39
Petrie nets, 18, 33
Pevsner, N., 190
Piepho, R.S., 191
Piloty, R.A., 33, 38, 39, 61
Pipelining, 178, 190, 224
Pizzarello, A., 20, 23
PL/1, 38
PMS notation, 36
Pointillism, 176
Policy decisions in architectural design, 227
Popper, K.R., 213
Portability of programs, 48
Portable C compiler, 201
Postcondition, 70, 74, 238
P ratio, 198–199

Precondition, 70, 73, 238
Private procedures in S*A, 51–55
Private variables in S*A, 46, 53–55
Procedure call function, 209–210, 222–231
Procedure deactivation in S*A, 47–48
Procedure invocation in S*A, 47–48, 96
Procedures in S*(QM-1), 161–165
Process construct, 38
Producer-consumer problem, 94
Programs, measurement and analysis of, 199–210
Proof outline, 72
Proofs of correctness:
 concurrent systems, 92–95, 98
 partial, 70, 92, 109
 sequential systems, 70–72, 84, 236–245
 total, 70, 109, 245
Pseudoconstants in S*A, 45
Pseudovariables:
 in S*, 131
 in S*(QM-1), 159
P type instructions, 197–199
PUMA Computer, 114

QA-1, 111, 125
QM-1, 9, 41, 49, 61, 114, 118–119, 125, 185, 216
 ALU, 144, 146
 architecture, 142–158
 control matrix, 156
 control store, 154
 data path organization, 142
 F store, 154–155
 Index ALU, 146–151
 K vector, 119, 143, 156–158
 leading edge operations, 157
 main store, 151–153
 microprogram counter, 153
 nanocontrol, 156–158
 nanostore addressing, 158
 priority address select mechanism, 158
 shifter, 146
 T step, 157
 T vector, 119, 143, 156–158
 trailing edge operation, 157
 as universal host machine, 171
QM-C, 187, 200
 methodology of design, 190, 216

Radin, G., 180
Ramamoorthy, C.V., 10, 125, 190
Randell, B., 21, 23, 185, 221
Rauscher, T.G., 10, 118, 122, 125
Reddi, S.S., 5, 10

INDEX 299

Reduced instruction set computers, 180, 191, 223
Reed, I.S., 24
Refutability, criterion of, 213
Region statement in S*, 137, 167–168
Register transfer level, 25, 33, 35, 38, 46, 48, 182–183
Repetition statements, 46–47
Repusse, 176
Residual control, 141, 154
Rideout, D.J., 114, 126, 172
RISC-I, 180, 187, 202
Ritchie, D.M., 201, 216
Robinson, L., 39
Rosen, S., 190
Rosin, R.F., 10, 154
Russell, E., 221
Russell, R.D., 9, 10

S*, 38, 40, 41, 61
S*A, 32, 37, 38, 39, 41, 49, 50, 61, 64, 226
SAL, 200
Salisbury, A.B., 10, 125
SARA methodology, 22, 23, 36
Satisficing solution, 42, 174
Semantic gap, 192–193
Semantics:
    of micro-operations, 117–118
    of programming languages, 29
    of S*, 260
    of S*A, 250
Semaphore, 56, 94
Sequin, C., 180, 187, 200, 202, 203, 209
S* family of languages, ix, 40–43
Shaman, P., 10
Shibayama, K., 125
Shriver, B.D., 38, 39, 121, 126
Side effects in assignment statements, 133
Siewiorek, D.P., 190, 191
SIMD machines, 111, 190
Simon, H.A., vii, ix, 1, 10, 42, 64, 174, 175, 190, 194, 212
SIMULA, 120
Simulation, 14, 21–22, 33, 87
Sint, M., 39, 125
SISD machines, 177
Sitton, W.G., 125
SLIDE, 32, 38, 39, 45, 61
Smoliar, S.W., 30, 38, 39
SODAS methodology, 22
Software engineering, viii
Specifications, operational and functional, 28–30, 39
Spillers, W.R., 17

SPL, 208
Split cycle, 153
S*(QM-1), 61, 141, 158–171, 231
Stallybrass, O., ix, 19, 23
Stcycle statement, 134–137
Steadman, P., 17
Stone, H.S., 10
Strecker, W.D., 191
Stritter, S., 221
STRUM, 38, 114, 119, 126
Student PL Machine, 52–55, 61
Style, ix, 23, 214
    architectural, 176–180
    in architecture design, 185–190
    in art, 175, 176, 190
    in artifacts, 174
    in building architecture, 175, 190
    in computer architecture, 174, 216, 221, 247
    in design, 176, 180–184
    in endoarchitecture, 177–179
    and esthetics, 175
    in exoarchitecture, 179–180
    functional, 177
    of implementation, 184–185, 196, 217, 224
    in manufacturing, 175
    in metallurgy, 176, 190
    microarchitectural, 180
    object oriented, 179
    pipeline, 178
    RISC, 180
    stack oriented, 179
    systolic, 184
    von Neumann, 37, 177, 219
Sumner, F.H., 17, 91
Suppe, F., 23
S*(V75), 139
SWARD, 57, 58, 61, 187, 224
Sykes, J.B., 23
Symbolic design, 41
Synchronization in S*A, 56–57, 87, 96
Synchronization mechanisms, 38, 93, 96
Synonym declarations, 68–69
Synonyms:
    in S*, 131–132, 160
    in S*A, 57–58, 235–236
Syntax:
    of S*, 260
    of S*A, 250
System construct in S*A, 42, 49, 50–55

Tagged data, 210
Tan, C.J., 125
Tanenbaum, A.S., 187, 191, 200, 203, 208, 209

Tartar, J., 114, 125, 141, 172
Termination, 109
Texas Instruments ASC, 178
THE operating system, 176, 195
Thomas, D.E., 191
Timing issues, 42, 87, 117–118, 134–137
   monophase, 118
   polyphase, 118
Tokoro, M., 125
Tomita, S., 125
TRANSLANG, 114
Tredennick, N., 221
Treleavan, P., viii, 39
Tsuchiya, M., 125
T vector, 119, 143
Two level microprogram store, 118, 140–141, 221

Universal host machine, 171
UNIX operating system, 201, 202

Varian 75, 117, 125, 139
VAX, 202
VENUS operating system, 176
Vertical migration, viii

Very large scale integration (VLSI), viii, 48, 117, 180, 184, 190, 191, 223
VMPL, 119
von Neumann machine model, 31

Wagner, A., 260
Wallace, J.J., 32, 38, 39, 61
WELLMADE design methodology, 20–21
Whyte, L.L., 10
Wilkes, M.V., 3, 10, 15, 17, 63, 64, 73, 84, 221
William of Occam, 81
Wirth, N., viii, 210, 215, 250
Wood, W.G., 126
Wortman, D.B., 61, 200–203, 210
Wulf, W.A., 209, 211

XPL, 20

Yau, S.S., 126

Zemanek, H., 10
Ziegler, S., 191
Zimmermann, G., 188, 191
Zurcher, W., 21, 23, 185